10/17

Rabbit

Rabbit

THE AUTOBIOGRAPHY OF MS. PAT

PATRICIA WILLIAMS

WITH JEANNINE AMBER

DEY ST.

An Imprint of WILLIAM MORROW

RABBIT. Copyright © 2017 by Patricia Williams and Jeannine Amber. All rights reserved. Printed in the United States of America. No part of this book may be used or reproduced in any manner whatsoever without written permission except in the case of brief quotations embodied in critical articles and reviews. For information, address HarperCollins Publishers, 195 Broadway, New York, NY 10007.

HarperCollins books may be purchased for educational, business, or sales promotional use. For information, please email the Special Markets Department at SPsales@harpercollins.com.

FIRST EDITION

Designed by Michelle Crowe

Library of Congress Cataloging-in-Publication Data has been applied for.

ISBN 978-0-06-240730-6

17 18 19 20 21   LSC   10 9 8 7 6 5 4 3 2 1

To my husband and our four kids, I love you all

Contents

CONTENTS

Life's a bitch. You've got to go out and kick ass.

—*Maya Angelou*

INTRODUCTION

W e'd been living in our new place in Indianapolis for only a couple of days when I heard a knock at my front door. I opened up to find a white lady with a big smile standing on my porch, holding a huge chocolate cake wrapped in plastic. "I want to welcome you to the neighborhood," she said. "So I baked you a little something."

What the hell? Where I'm from, if somebody shows up at your door with something nice in their hands, it's probably stolen.

As soon as she left, I went right to my kitchen and called my girlfriend, Ms. Jeanne, back home in Atlanta.

"You ain't gonna believe this," I said. "A white lady just made me a cake. You think I should I eat it?"

"Yeah, girl," said Ms. Jeanne. "White folks always bake you shit when you move in so you don't break in to their house."

It turns out, there was a whole lot I had to get used to moving from the hood to the suburbs. Strangers bringing me chocolate cake was only the beginning.

I grew up in the 1980s in the inner city of Atlanta. My mama was an alcoholic single mother with five kids. She could barely read and only knew enough math to play the numbers and count out the exact change to buy herself a couple of bottles of Schlitz Malt Liquor and a nickel bag of weed. Almost none of my relatives, going back three generations, ever graduated high school. Instead, you could say I came from a family of self-employed entrepreneurs. My granddaddy ran a bootleg house, selling moonshine out of his living room; my uncle Skeet robbed folks; and my aunt Vanessa sold her food stamps. With role models like that, what could possibly go wrong?

Even though I came up in the hood, I dreamed of a different life. My fantasy came straight off TV, from my favorite show, *Leave It to Beaver*. You probably thought I was going to say *Good Times*, but I didn't need to watch TV to see black folks struggling. The Struggle was all around me. Compared to how we were living, life on *Leave It to Beaver* looked like heaven. I was mesmerized by the way the house was so clean and everybody was always smiling and jolly. What I liked most was how Mrs. Cleaver would walk around grinning at her kids like she couldn't believe her good luck. In my house, my mother would get drunk off her gin, whoop me with an extension cord, call me ugly, and tell me to take my ass to bed. I'd be thinking, *How you gonna tell me to go to sleep when it's ten o'clock in the morning and I just woke up?*

I know a lot of people think they know what it's like to grow up in the hood. Like maybe they watched a couple of seasons of *The Wire* and think they got the shit all figured out. But TV doesn't tell the whole story. It doesn't show what it's like for girls like me; how one thing can lead to another so that one minute you're a twelve-year-old looking for attention, then suddenly you end up pregnant at thirteen, with nobody to turn to for help. Folks don't know about that kind of life because, for a lot of people, girls who grew up like me are invisible. Unless you come to the hood, you won't see us. It's easy to pretend we don't exist.

By the time I was fifteen, I was a single teen mom with a seventh-grade education, no job skills, no money, and two babies under the age of two. My dream was to give my kids a better life, but most days I didn't even have enough money to buy Pampers. All I wanted was to find a way to get myself and my babies out of the ghetto; I was willing to do whatever it took.

Let me tell you something, moving up in this world is not easy. I worked at factories, gas stations, and fast food restaurants. I've hustled and schemed, been shot twice, beaten with a roller skate, locked behind bars with a bunch of junkies and hookers, and nearly got my head blown off for talking shit. Somehow, I survived. Hell, I did *more* than survive. I got myself and my kids a whole new life.

These days I live with my family in Indianapolis, in a six-bedroom house overlooking a man-made pond with a bunch of ducks swimming around in it. During the day, I do regular suburban-mom-type shit. I go to Walmart, get some lunch at Chick-fil-A, and head over to the gym for Zumba class. Okay, I don't really do Zumba. I went once, but the teacher was plus size, like me. I kept thinking, *Does this shit even work?*

At night I hit the clubs. I'm a comic and tour the country telling stories about my messed-up childhood and getting out of the hood. When I started comedy, back in 2004, all I wanted was to make folks laugh. Then I noticed something strange. After almost every show somebody would come up to me and ask the same question, "How did you turn your life around?" It felt like they wanted me to give them some kind of secret tip.

I wish I had a simple answer. But the truth is, it's a long story. I went from living in an illegal liquor house, to running from the cops, to living in the suburbs with a flock of ducks outside my window. The only way I can explain *how* it happened is to tell you exactly what went down. So I'm laying it all out in black and white, sharing stories I've never told a soul, not even my husband, which reminds me, I should probably warn him about chapter 5.

I used to get embarrassed about the shit I did to survive. I wanted to push it all away and pretend it never happened. But I've learned that laughing at my pain helps me heal. I hope my story will inspire you to laugh through your hard times or try something you've always dreamed of doing. Maybe you want to get out of a bad relationship, or go back to school, or change your career. Hell, maybe you want to be an overweight Zumba instructor. I don't know what the hell you lie in bed thinking about at night. That's your business. All I know is when you finish reading this book I hope you'll take away the same message that I've been carrying in my heart since I was eight years old. It's a lesson an angel taught me. That angel happened to be my third-grade teacher, who wore badass leather boots and had really good hair. The words she spoke to me all those years ago helped me change my life, and maybe they'll do some good for you, too. "Patricia," she said, "I want you to always remember, you can do anything and be anything. All you have to do is dream."

CHAPTER 1

Bear Cat

My granddaddy is the only black man I've ever met who was never broke a day in his life. He ran an illegal liquor house in Decatur, Georgia, selling moonshine for fifty cents a shot from behind a bar he built himself out of plywood and old scraps of carpet and red leather. Granddaddy's real name was George Walker, but folks called him Bear Cat or .38 for the two pistols he kept in his front pockets. Granddaddy didn't believe in banks and didn't trust anybody, either. He stored his jugs of corn liquor in the living room in a beat-up old refrigerator the color of baby-shit yellow, which he locked up with a thick metal chain. And he stashed his money in a dingy white athletic sock he pinned to the inside of his pants. My brother Dre, who would steal anything that wasn't nailed down, used to say he'd be one rich muthafucka if he could only get his hands on that sock full of paper. But Dre didn't want to swipe anything that hung so close to Granddaddy's mangy old balls.

Most folks were scared to death of my grandfather, not just because he was built like somebody put a human head on a gorilla body, but also because he didn't take shit from anybody. I remember one night my uncle Skeet was acting a fool while Granddaddy was trying to watch Walter Cronkite on the evening news. The news was serious business to Granddaddy. He liked to talk back to Mr. Cronkite like the two of them were having a real conversation: "What's wrong with these dumb-ass honkeys?" he'd yell at the TV. "They finna elect a movie star to run this whole gotdamn country. This why a nigga don't vote!" Or, "Them Iranians some mean muthafuckas. That's why I don't go nowhere!" Granddaddy said other than Jesus Christ, Walter Cronkite was the only white man he could trust. Yet here was Uncle Skeet, drunk as Cooter Brown, bouncing on the balls of his feet and shadowboxing right in Granddaddy's face in the middle of the news.

Granddaddy waited till the commercial break, then he grabbed an old golf club he kept behind his bar and smashed Uncle Skeet right across the jaw, knocking out his front teeth. When the news came back, Granddaddy stopped swinging and sat back down in front of his little black-and-white set, cool as a cucumber, like nothing happened. After that, when the news came on, nobody made a sound.

Back then there were nine of us living with Granddaddy in his big yellow house on Arkwright Place: me, my mama Mildred, Mama's boyfriend Curtis, my sister Sweetie, and my three brothers. Also, Uncle Skeet who broke into houses and stole shit for a living, and Uncle Stanley who was crippled and slow in the head and had to go to a special-needs school. The bedrooms were in the back of the house and the bar was in the living room, up front. Granddaddy had decorated it with old bedsheets nailed above the windows like curtains, and pictures of Martin Luther King Jr. and Jesus Christ hanging on the wall. The main difference between a regular bar and a bootleg house is that a regular place closes at night and everybody goes home. At Granddaddy's, folks drank, played spades, shot craps, and hollered at each other until they passed out. On the weekend, it was like a sleepover with the neighborhood drunks. I hated all the noise and commotion. At night I'd go to sleep hoping that I'd wake up and find myself magically living in a clean house where nobody punched each other, no matter how mad they got. But instead I'd get up and find some stranger passed out cold on the living room floor, covered in their own piss and puke. That's the mess I grew up in. When I was six years old, I thought everybody lived that way.

"MILDRED BABY GIRL!" Granddaddy called for me one morning, his voice booming through the house. Mama had five children, to this day I am not sure if Granddaddy knew any of our real names. When he wanted us, he'd call us by the order we were born. "Mildred First Boy!" was my oldest brother, Jeffro; "Mildred Baby Girl!" was me. Everybody knew I was Granddaddy's favorite. When he hollered, I'd come running.

"Help me fix these grits," he said, when I found him in the kitchen that morning. He was holding a thick metal chain in his hands, because the same way Granddaddy kept his moonshine and guns locked up tight in a fridge in the living room, he also padlocked the fridge in the kitchen. Other kids knew it was mealtime when their mama called them to the table. We knew we were gonna eat when we heard that chain hit the floor.

Granddaddy pushed a chair to the stove and lifted me up so I could stir the pot while he fried up eggs and fatback in the pan beside me. "That's real good," he said, looking over my shoulder. "Baby girl, you a natural in the kitchen, musta got it from me."

Granddaddy's specialty was homemade cat head biscuits, which were the biggest, fluffiest biscuits you could ever eat, and came the size of an actual cat's head. He also cooked chicken back, which is 90 percent skin and bones, except for the piece at the end that covers the chicken's asshole. That piece is 100 percent fat. Granddaddy would cook it in the skillet, drain it on some newspaper, and set it on the table with a bottle of Trappy's hot sauce. Sometimes I'd pick up a piece of chicken and it would have the news of the day printed all over it.

Everybody used to joke that I stayed up under Granddaddy like a baby chick to a hen, holding onto his pant leg and following him around wherever he went. It's true. I loved that man with every inch of my whole little heart. Granddaddy made me feel safe. But my mama—she was a whole different story.

"MOVE OUT THE WAY so the kids can cut a rug!" Mama hollered, pushing me and my sister Sweetie into the middle of the living room. It was Saturday night and the place was jumping. Anita Ward was singing about somebody ringing her bell on Granddaddy's little record player, while Mama, drunk as a skunk, yelled for everybody to clear the floor so her two little girls could dance.

Mama was an alcoholic. She drank Schlitz Malt Liquor and Seagram's Extra Dry Gin, which she called Bumpy Face because of the bumpy texture of the glass bottle. Mama's drinking was the main reason she didn't act like any of the mothers I saw on TV. She didn't help with homework or give us kids advice. She didn't care about bedtimes, or even where we slept. There weren't enough mattresses for all the people who lived at the liquor house and it was nothing for Mama to stumble over one of her children sleeping on the floor. She'd just step right over us and keep moving. I don't remember ever hearing Mama say, "I love you" or "You did good." In fact, she barely took the time to name her own kids. I have three brothers; one is named Andre and another is named Dre. That's the same gotdamn name, and those two aren't even twins.

In the living room, Mama turned up the music.

You can ring my beeeeeell, ring my bell
Ring my bell, ring-a-ling-a-ling

"C'mon now," she said, pushing folks out of the way. "Let the babies dance!" My sister Sweetie loved the way everybody was looking at her and started shaking her little ass with a big smile on her face. But I hated the music pounding in my ears and all those eyeballs watching me. There was no way I was gonna let loose and get on down. Instead, I did the two-step with my face fixed like I was sucking on a lemon. But it didn't even matter, those drunk-asses still enjoyed the show. Sitting in a beat-up old chair by the window, Mr. Tommy, a regular, leaned back to watch my sister. He looked at her like she was a juicy piece of chicken and he was about to dig in. "Mmmm-mmm," I heard him say to his brother, Po Boy. "She look *real* good." Sweetie was eight years old.

I hated when Mama made us dance, but she did it all the time. I never knew why until one night when I saw Mr. Tommy slip her a couple of dollars right before she pushed me and my sister onto the floor.

MAMA WOULD DO ANYTHING for a little extra cash. Anything, that is, except get a regular job. Her big moneymaking scheme, the one she came up with when I was seven years old, was picking pockets. Only she didn't want to do the dirty work herself. Instead, she'd wake me up in the middle of the night and make me do it for her. I guess that was her way of giving me on-the-job training.

"Rabbit!" I heard her call the first time. I was asleep on a blanket on the floor in the bedroom Mama shared with her boyfriend, Curtis. Sweetie was beside me, curled up in a ball.

"RABBIT!"

I opened one eye and saw Mama standing over me. "Get your ass up," she hissed, waving at me to follow her. She led me to the entrance of the living room and pointed inside. "See that?" she said. "They out cold." The room was filled with leftover drunks from the night before. Mr. Tommy was asleep in a raggedy armchair by the bar with Po Boy knocked out beside him. Our neighbor Miss Betty was laid out, barefoot, on the sofa with her wig sliding off her head. In a chair by the card table was Mr. Jackson, the janitor from my brothers' school, his head back and mouth hanging open.

Mama nodded toward Po Boy: "Go in there and pinch his wallet."

"Huh?" I asked, confused.

"Take his wallet out his pocket and bring it to me. I'll give you a dollar."

I looked at Po Boy, then back at Mama. "What if he wakes up?"

"Chile, he ain't waking up." Mama took a step toward Po Boy and waved her hands in front of his face. "See?" she said. "He asleep."

I stared at Po Boy; he had a thin stream of drool running from his mouth. Mama reached over and shoved him on the shoulder. His head fell forward, then jerked back. She nudged him again and he still didn't move. "I told you he ain't gonna wake up," Mama said, satisfied.

What I didn't understand was why she didn't pinch the wallet herself. She was already standing right there, pushing and poking the man. What did she need me for? But I didn't say a word. As scared as I was that Po Boy would suddenly open his eyes, find me digging for his wallet, and whoop my ass, I was even more afraid of Mama. One time she told me to get her a cup of tap water to chase back her gin and I didn't move fast enough. So she made me bring her three switches from the yard and soak them in the tub. Then she braided them together and beat the dog shit out of me.

"Go on," said Mama, pushing me toward Po Boy. "Go on and get it."

Po Boy's overcoat was hanging off his shoulders, making a puddle of cloth on the floor. I held my breath as I felt around for an open pocket and reached inside. When my hand touched the smooth leather of his wallet, I grabbed it and ran back to Mama, who was waiting in the doorway—I guess so she could make a break for it if Po Boy suddenly woke up.

She opened the wallet, took out a wad of bills and shoved them in her bra.

"Where's my dollar?" I asked, holding out my hand.

Mama's eyes got real squinty. She took the stolen money out of her bra, peeled off a single dollar bill, and held it out to me. When I went to grab it, she hung on to it a second longer than she needed to.

"Listen," she said, real slow. "Go put this wallet back in Po Boy's pocket. Then go get the wallet from Mr. Jackson. Do it quick, before he wakes up. I'll give you another dollar."

That was the first time Mama made me steal. But I knew by the look on her face and the money in her bra, that she was going to make this a regular thing. Sure enough, from then on, almost every Sunday morning before the sun came up, Mama would kick me awake so I could help with her crime spree.

The upside was that with all those blackout drunks, I was making *good* money—five dollars was a lot for a kid in 1980—and I spent it all at the corner store. I wasn't stingy, either. I treated my brothers, sister, and cousin to all-they-could-eat Laffy Taffy, Hubba Bubba, and Pop Rocks. And I played so much Pac-Man that my name *stayed* at the top of the scoreboard: R-A-B for Rabbit, which is the name Mama's boyfriend Curtis gave me when he came home one day and found me sitting on the porch eating a carrot.

But as much as I liked the money and respect, deep down I hated my job. My stomach went in knots every time Mama made me sneak my hand into somebody's pocket. I wanted to tell her, "I'm a little kid. I don't have the nerves for this!" Even Curtis tried to get Mama to stop. "This ain't right," I heard him tell her one night. "All you doing is teaching your little girl to steal."

But Mama didn't care. To her, there was no such thing as a bad hustle. It was all good, just as long as you didn't get caught.

CHAPTER 2

Hot Lead

I was out in the yard catching fireflies and holding them up to my face pretending they were earrings, when I heard Granddaddy call my name. "Mildred Baby Girl!" he hollered. It was Saturday night, and he was calling me to come watch our all-time favorite TV show: *Georgia Championship Wrestling.*

"Baby Girl!"

I dropped everything and flew up the front steps to the screen door. Granddaddy had thrown out so many customers without opening it first that the mesh screen hung off the doorframe like a skirt flapping on a clothesline. I pulled the screen aside, and ran indoors.

It was still early, but the living room was full of people. The Numbers Man was sound asleep in a chair by the window with his belly resting in his lap like a giant egg; Mama and Auntie Vanessa were on the sofa sipping on corn liquor and arguing about who got the better voice, Lou Rawls or B. B. King; Mr. Tommy and his brother were sitting at the card table in the corner, playing spades. Granddaddy was behind the bar, waiting on me.

I whipped through the room and hopped on my stool just as he was switching on the little TV that he had sitting on the bar. The set flickered and Granddaddy turned to me, "You ready to see some ass whooping, Baby Girl?"

"You know it!"

Granddaddy's favorite wrestler, Claude "Thunderbolt" Patterson, was in the ring, crouched down low and strutting around like he was doing the funky chicken. "That's *my* man," Granddaddy said. We leaned in close and watched Thunderbolt charge at his opponent, head-butting him so hard he fell to the mat like an old wet rag. "Look at Thunderbolt whoop that cracker's ass!" yelled Granddaddy.

"Are they fighting for real?" I asked, bouncing on my stool.

"Of course it's real," he answered. "You see how that cracker's laid out? Ooooweee! Thunderbolt put a hurtin' on his ass." Granddaddy gave me a sideways smile, then slowly backed up off his stool with his fists up, like the two of us were gonna fight. I jumped down and squared off against him, weaving from side to side, with my scowl face on. The two of us practiced our wrestling moves every Saturday night; it was the best part of my week.

"What you got for me, Baby Girl?" Granddaddy growled. "What you got?"

I wound up my arm and took a step back.

"You know what's coming for you!" I hollered, and ran toward him.

Granddaddy caught me in his giant paws and threw me over his shoulder. "I got you now!" he yelled. "I got you!" Giggling like crazy, I tried to get him in an upside-down headlock while he spun me around. Out of the corner of my eye, I could see everybody pass by in a blur: Mr. Tommy, the Numbers Man, Aunt Vanessa, and Mama in the corner. And then I saw Miss Betty. She was walking through the door, wearing a faded red dress that was buttoned wrong so that the hem hung uneven. On her head was some hair that looked like it might have originally been a good Sunday wig, only it was raggedy as hell and sitting way back, so her tore-up edges were on full display. Miss Betty stepped up to the bar just as Granddaddy was putting me down.

"Bear Cat, lemme get a drink!" she demanded, slapping her hand down on the counter.

To this day, I don't know what she was thinking interrupting us during wrestling. Everybody knew not to bother Granddaddy when he was watching TV. But here she was with her half a wig, right in the middle of our show. "C'mon, George," she said. "I need a drink."

Granddaddy didn't even look away from the set. "Where your money at?" he asked.

"I'll pay you tomorrow," she said. "You know I'm good for it."

It's true that Granddaddy sometimes let his regulars get drunk on credit. Like if they were teachers or construction workers, folks with regular jobs and steady paychecks. But Miss Betty didn't have any kind of job. She was a full-time drunk and sometimes ho, which I knew for a fact because once a month Granddaddy would pay her twenty dollars to go in a bedroom in the back and fuck Uncle Stanley.

Uncle Stanley had something wrong with his legs, sort of like they were nailed shut at the knees. When it was time for him to get it on, Granddaddy would call for me and Sweetie to help. "Go on back and help your uncle get started," he'd say.

Uncle Stanley would pull himself on top of Miss Betty, then I would take one leg and Sweetie would take the other and we'd yank them apart. As soon as he got a steady rhythm going we knew he was good to go and we'd run outside to play. I guess Granddaddy thought he was doing some good parenting by helping his disabled son get some pussy. Who cared if it was from a broke-down ho with two little girls holding down his legs?

For a minute Miss Betty just stood at the bar looking stupid, while Granddaddy ignored her. Then she started to yell: "Fuck you, Bear Cat! You ain't nothing but a big black faggot!" The whole room suddenly got quiet. Even *I* knew Miss Betty had messed up. You don't call an old black southern man a faggot unless you're ready to be carried by six. Granddaddy jumped off his stool, grabbed Miss Betty by the arm, and pulled her out the front door. She stumbled down the steps and fell into the dusty yard, still hollering and cursing.

"Get the fuck outta here," Granddaddy called from the porch. "You ugly-ass bitch." I had come running out after him and stood beside him with my eyeballs bugging out of my head, watching the action like I was at the movies.

Miss Betty got to her feet and pointed her finger at Granddaddy. "Go to hell and kiss my muthafuckin' ass!" she hollered. Then she turned around, bent over, and slapped her behind. "Kiss it, you gotdamn faggot!"

That's when he shot her.

The first bullet hit Miss Betty in her left butt cheek. She spun around and he shot her again. This time he blew off her pinkie finger. She fell to the ground screaming, but Granddaddy just kept on firing like he was at a shooting range.

"Lord have mercy!" cried Aunt Vanessa, running onto the porch. "Daddy, what'd you shoot that lady for?"

"Fuck her," Granddaddy answered. Then he added, turning to my aunt. "'Nessa, go on inside and pour the liquor down the drain."

Auntie Vanessa just stood there.

"Girl, go do what I told you!" he yelled. "Hurry up and get rid of the shit. Then call the police."

To this day, I don't know why my Granddaddy was more worried about getting caught for moonshine than attempted murder. Maybe it's because he thought Miss Betty had it coming. "I shot her," he said matter-of-factly when the police showed up. "I gave that bitch some hot lead." The cops shook their heads, put my grandfather in the back of their patrol car, and took his ass to jail.

Granddaddy got locked up for a lot of years behind that mess. Once he was gone, there was no one to run the liquor house and we all had to leave. Aunt Vanessa took in Uncle Stanley; Uncle Skeet got busted for burglary and ended up in jail. For a while Mama and us kids lived with her boyfriend, Curtis, in a three-bedroom house with a chicken coop in the back yard. Curtis took care of us, paying the rent and keeping us fed. But it wasn't long before Mama ran him off by drinking too much and acting too crazy. Then it was Mama all by her lonesome, drinking her gin and struggling to take care of all us kids by herself.

When we lived in the liquor house, I used to hate the noise and commotion and the smell of stale cigarette smoke that never went away. I hated waking up to strangers in the living room and stealing from folks in the middle of the night. But I didn't realize that compared to what came next, that shit-hole bootleg house with the bedsheets in the windows and drunks passed out on the floor really wasn't all that bad. At least at Granddaddy's I always had food to eat, a roof over my head, and somebody who loved me. After he went to jail, and Curtis left us, all I had was Mama. In other words, I was eight years old and I was pretty much fucked.

CHAPTER 3

Struggling and Scheming

P atricia!" snapped Miss Thompson, looking at me with her face twisted up like she smelled some dog doo-doo. "You're *very* tardy."

It was 8:45 A.M. and I'd just walked into my third-grade classroom. Miss Thompson was doing her best to make me feel bad for showing up late. But she didn't need to. I had my own reasons for wanting to get to school on time. If I wasn't there by 8:00 A.M., I missed getting Free Breakfast. That little box of Kellogg's Corn Flakes and itty-bitty container of apple juice were the only things I liked about school. I hated being late more than Miss Thompson could ever imagine. But I couldn't get to school on time when nobody woke me up.

It had been more than a year since we'd left the liquor house, and Mama, Sweetie, my brothers, and I were living in a run-down two-bedroom duplex on Griffin Street, on Atlanta's West Side, across the street from a family who kept a dirty brown sofa and busted refrigerator in their front yard. Mama didn't see a reason to open her eyes until *The Price Is Right* came on at 11 A.M. So it was usually Dre who woke me up in the morning by kicking me in the leg and hollering, "Git up, girl!" But he wasn't the most reliable.

Dre was eleven years old and "living his life," as he liked to say, which meant he was busy stealing college kids' bikes off the campus at Georgia Tech. Sometimes he'd get caught by the popo and thrown into juvenile detention. When that happened, nobody would wake me up and I'd come to school late.

Miss Thompson stared at me standing in the classroom doorway, sighed, and rolled her eyes. "All right," she said, finally. "Go hang up your coat and come take your seat."

"Yes, ma'am."

I could feel the eyeballs of every kid in the class on my back as I walked across the room to the closet behind the blackboard. I hated those kids almost as much as I hated being late.

Porsha and Mercedes were my biggest enemies. I don't know what I ever did to make them mad, but those two little bitches made it their mission to make my life miserable. After school they'd roll up on me and Sweetie and tease the fuck outta us. "Nasty-ass bitches!" they'd scream. "Look at your nappy-ass hair! Your shoes is raggedy! You nasty and you *stink*! Y'all smell like dog shit." I guess being named after luxury vehicles made them feel like they were better than everybody.

"Fuck you, Pontiac!" I'd holler at Porsha, and she'd damn near lose her mind. "It's POR-SHA!" she'd yell.

In the closet, I slid off my jacket and hung it on my hook. My coat was red with a dingy used-to-be-white collar and dirty cuffs. I leaned over and gave it a sniff. In fact, it did *not* smell like dog shit. But it sure did smell. The odor was more a mixture of wet mildew and dried funk. It wasn't even my funk, either. I don't know who that jacket used to belong to. Dre and Jeffro had stolen it one night, along with a bunch of other clothes, from the donation bins out behind the Goodwill store on North Avenue. Most poor folks shopped *inside* the Goodwill, but we were so broke my brothers had to rob the place.

As much as I hated school, I liked being alone in the coat closet. It was cozy and quiet. I ran my hands over the different-colored jackets hanging on their hooks. Porsha had a light pink coat with pretty fake fur around the hood, so I stuck my index finger in my nose and wiped it on her collar. "There's your present, bitch," I whispered.

Above the coat hooks was a shelf where all the kids who didn't get Free Lunch kept their lunch boxes. They were filled with every kind of sandwich you could think of—bologna, peanut butter and jelly, souse meat, sliced ham, government cheese—all of them cut in triangles and wrapped in tinfoil. At lunchtime kids would sit in the cafeteria and lay out their sandwiches, thermoses filled with Kool-Aid, and little sacks of Lay's potato chips, like they were at a swap meet. All by myself in the closet, surrounded by all that food, I could hear my empty belly calling out to me. "Girl," it said, "I'm hungry!"

I had missed Free Breakfast, and Free Lunch was hours away. But it occurred to me that Mercedes, who was fat as hell anyway, probably wouldn't notice if I ripped off an itty-bitty piece of whatever sandwich her mama had packed in her lunchbox. Hell, sometimes that heifer had *two* sandwiches. There was no way she'd notice if a little corner was missing. I grabbed her blue Smurf lunch box off the shelf and crouched down under the coats. When I opened the lid, the smell hit me like some good perfume: ham and American cheese on soft white bread, dripping with Miracle Whip.

We didn't have this kind of top-shelf food at home. Mama used her food stamps to buy runny no-name ketchup and cheap Sunbeam bread. Put those together, you got a ketchup sandwich, also known as dinner. One time Mama came home with a big tin can of government peanut butter. You knew it was government because it said "surplus" on it. That's not a brand they sell at the Super Saver. That peanut butter was a health hazard. It was dry as hell and would get stuck in your throat like a ball of concrete. After the time Andre almost choked to death on a sandwich, we learned not to eat the government peanut butter unless you had a big cup of water right there ready to wash it down.

I pulled off a tiny piece of Mercedes's sandwich and put it in my mouth. But I was so hungry, and she was so fat, I thought, *Fuck it,* and started shoving that delicious sandwich in my mouth faster than I could swallow. I closed my eyes and let out a little moan. It was like I'd died and gone to sandwich heaven. I was so deep in my feelings of enjoyment for this good-good food, I didn't even hear Miss Thompson come into the closet. I opened my eyes and there she was, standing right beside me.

"Patricia Williams!"

"Yeth?" I said, looking up and swallowing hard.

"Young lady, are you back here eating up somebody else's lunch?"

"No, ma'am."

"Then whose is that?" she asked, pointing at Mercedes's blue lunch box in my lap.

"I don't know," I answered. Miss Thompson glared at me with her eyeballs popping out of her head.

"Why?" I asked, holding up the sandwich. "You want some?"

Miss Thompson didn't answer. Instead she grabbed me by the top of my arm and yanked me up. "Honest to God!" she cried, "I've seen better behavior in a barn full of animals." Then she dragged me out of the classroom and across the hall to the principal's office. "We'll see what Mr. Dixon has to say about this."

I'm pretty sure Miss Thompson had been waiting for an excuse to take me to the principal's office ever since I landed in her third-grade class. That lady *never* liked me. One time I came to school looking extra raggedy, not only because of my Goodwill outfit of a faded yellow T-shirt and high-water jeans, but also because the night before Mama had decided to take out my hair and rebraid it. Only she passed out when she was halfway done. I showed up for school the next morning looking like I was wearing one of those half-man, half-woman costumes, except on one side I was Buckwheat from *The Little Rascals,* on the other side I was Penny from *Good Times.* When Miss Thompson saw me walk into her classroom she just stared with her hand covering her open mouth like she'd never witnessed this kind of child abuse. At first I thought she felt bad for me, but when Porsha came in behind me, Miss Thompson said nice and loud, "Don't you look pretty as a picture, Porsha, It's nice to see your mama takes such good care of your . . . *grooming.*" Then she gave me the side eye like it was *my* fault Mama was a drunk.

In the principal's office, Miss Thompson went behind the counter and talked to the secretary. The two of them kept looking my way and shaking their heads. Then Miss Thompson went back to her classroom, leaving me sitting on the wooden bench swinging my feet and waiting on Mr. Dixon.

Maybe if he'd asked me why I took the sandwich, things would have been different. Maybe if I'd had a chance to tell Mr. Dixon I was hungry, that I missed Free Breakfast, that Mama didn't cook anything the night before and all I had for dinner was a few bites of the Jumbo Honey Bun Dre had stolen from the corner store and split three ways with me and Sweetie, things might have turned out another way. But the minute I stepped into his office, it was obvious Mr. Dixon didn't give a shit about my empty belly. He just wanted to teach me a lesson.

He stared at me from behind his desk. "I'm very disappointed in your behavior," he said. "This is a very serious offense."

"Yes, sir."

"It's important that you understand that stealing will not be tolerated."

"Okay."

"Not tolerated at all."

To make sure I was "receiving the message loud and clear," he told me to stand up, put my hands on his desk, and bend over. Then he took his big wooden paddle and whooped my behind.

"Young lady . . ." *SMACK*

"At this school . . ." *SMACK*

"We . . ." *SMACK*

"Do not . . ." *SMACK*

"Steal!" *SMACK SMACK SMACK*

I suppose Mr. Dixon thought this was important information I needed to succeed in life. But bent over his desk with my ass on fire, all I could think about was food.

YOU LEARN TO LIVE with a lot of bullshit when you're poor as hell: cockroaches crawling on your toothbrush, no running hot water for a bath, having to pack up all your belongings in trash bags every few months and move because your mama fell behind on the rent. But the one thing I could never get used to was being hungry.

After we left the liquor house and moved to the duplex on Griffin Street, Mama had to take care of us on her own. She didn't have a job. Instead, she made do on a few hundred dollars in welfare and food stamps every month. But it was never enough. First the gas got cut off, then the electricity. One afternoon Mama made me go across the yard, to the apartment next door, and ask the old lady who lived there if I could plug an extension cord into her electric. That lady looked at me like she didn't understand what I was asking for. "My mama need it for the TV," I said, so she'd see it was an emergency. "*The Young and the Restless* about to come on." The old lady shook her head, then took the cord from my hand, ran it through her side window and plugged it in. But we still didn't have any lights. When night fell, we lit a pack of birthday candles and stuck them on the kitchen counter like itty-bitty table lamps. A birthday candle burns for about seven minutes, in case you're wondering. That's just enough time to make yourself a no-name ketchup sandwich and take your ass to bed.

Most nights I'd fall asleep thinking about food: Granddaddy's homemade grits simmering on the stove; the chewy pizza they served every Friday for Free Lunch; Mercedes's mama's sandwiches dripping with Miracle Whip. I even dreamed about food I'd never had before, like McDonald's, which apparently was too high class for us.

"Niggas don't eat McDonald's," Mama said every time a Big Mac commercial came on TV teasing me with pictures of mouthwatering burgers that I could never have. One morning I was watching Mama's broke-down set and the screen filled with a close-up of a Quarter Pounder. It looked so good, with the drops of water glistening on crispy lettuce leaves, I ran across the room with my mouth open to lick the TV, and damn near electrocuted myself.

As hungry as we were, I can't say Mama didn't try. She came up with all kinds of schemes to get us fed. One time she took us to the Curb Market late at night so we could dig through the dumpsters out back looking for anything the vendors had thrown out that wasn't too rotten to eat. We found an old head of cabbage, some wilted carrots, and a couple of stale loaves of bread. Another time she took me and Sweetie out with her for hours in the blazing sun to collect aluminum cans from the trash people put out by the side of the road. When we were done, we took our trash bags full of cans to Davis Recycling across town. For all our hard work, all we made was fifteen dollars and forty-five cents. It was just enough for Mama to go to the corner store and buy some necessities: two packs of Winston, three quarts of Schlitz Malt Liquor, a package of bologna, a loaf of Sunbeam, a box of Saltine crackers, and two cans of Libby's potted meat—which is a bunch of cow parts nobody wants to eat ground up to look like the kind of mess you find in your toddler's diaper when the baby has the flu. The food, the beer, and almost all the cigarettes were gone by the next day.

Mama was sure the recycling man was ripping her off, so she decided to get even. "Lookie here," she said to me and Sweetie the next week when we were out collecting cans. "Make sure you put some dirt inside."

"What you mean?" I asked, standing on the side of the road with an empty Colt 45 beer can in my hand.

"Pick up some dirt and put it inside the can," Mama repeated, like this was a regular everyday activity that any idiot would know. "Make sure you stomp it down and close it up so the dirt don't fall out." Mama figured this little trick would make the cans weigh more, and we'd get more money. But when we handed our trash bags over to the recycling man to get weighed, he picked up a bag and a pile of dirt spilled out. It looked like we'd spent the day at the beach.

"Lady," he said to Mama, "you can't pull that shit over here. I'll pay you for these, but don't come back here no more." He weighed the cans, then subtracted 30 percent for the dirt and handed Mama eleven dollars and thirty-five cents.

"Damn crackers trying to keep us down," Mama muttered to herself as we left. "Don't matter how hard you work, no one gon' give you a chance."

By the time we got home that afternoon, we looked something sorry. We were sweaty, hungry, and covered in dirt. Our neighbor Miss Cynthia, who was sitting on her stoop, must have sensed the desperation because she called out to Mama, "Y'all going to church tomorrow, Mildred?"

"No, ma'am," Mama answered. "I do my talking to Jesus at home."

"You know," said Miss Cynthia, "they got a real nice food pantry over there. They like to help folks out. Sometimes they even give you a little help with the rent. I'm just sayin', with you taking care of all those children by yourself . . . Hell, I been over there to get a few things from the pantry from time to time, and I *got* a man."

Mama told me and Sweetie to go on inside, then she sat down with Miss Cynthia to get the lowdown on all this food the church folks were giving away for free. I hate to say Mama was scheming on the Lord, but before she found out about the food pantry the only religion I heard about at home was when Mama told me I better pray to Sweet Baby Jesus she don't whoop the black off my ass. The day after she talked to Miss Cynthia was a whole different story. Mama was up bright and early, telling all us kids to put on some gotdamn clothes so we could go to muthafuckin' church.

Greater Springfield Baptist, a big red-brick building with white columns, was catercorner from where we lived. Every Sunday we'd see folks heading to worship: ladies in their good wigs, girls in pretty dresses, boys with hard-soled shoes and fresh haircuts. If Mama had known about the free food giveaway, we'd have found religion as soon as we moved in.

Mama told us to get dressed that morning, but she didn't say "nice." So we rolled into church looking raggedy as hell. We slid into the back row, Sweetie with her dirty flip-flops hanging off her feet and me in cutoff jean shorts and a pale green boy's T-shirt decorated with a picture of the Incredible Hulk. I looked around and quickly discovered that this was not regular church attire. The lady beside me was decked out in a purple dress, matching purple shoes, and a big-ass purple hat with all kinds of feathers and leaves and flowers decorating the top of it. It looked like she was wearing a gift basket on her head.

At the pulpit the preacher, dressed in a brown three-piece suit and dripping sweat, was yelling at the congregation like we were a bunch of bad-ass kids. "God tells us in his Fifth Commandment to honor thy mother and thy father!" he hollered. "I said HONOR thy mother and father!" He held his Bible high over his head, like it was raining out and he didn't want his hairstyle to get wet. "Now, brothers and sisters, what does it mean to honor your mother and father? Does it mean you gon' run out and buy your mama a brand-new color TV set?"

Beside me the lady in purple started to laugh, her bosom, covered in baby powder, jiggling like a bowl of Jell-O. "Oh Lord, I'd like that!"

"NO!" yelled the preacher, his voice bouncing off the walls of the church and right into my eardrums. "The true meaning of honor has nothing to do with the giving and receiving of material goods. It's about respect and obedience! Do you hear me? Respect and obedience. O-B-D-ence."

It felt like hours that I sat there watching the pastor wave his Bible in the air and yell. After a while, I looked over and was stunned to see that my three brothers were all knocked the fuck out, sound asleep with their eyes closed and heads rolled to the side. *How can they sleep through this scary-ass shit?* I wondered. The preacher was hollaring about eternal damnation and the white-hot fiery furnace of hell. I was more terrified than the time I watched *The Amityville Horror* on Mama's little black-and-white TV.

When the preacher was done yelling, everybody except my sleeping-ass brothers stood up to sing. *His Eye Is On the Sparrow.* Even Mama knew the words. And then—Praise Jesus!—it was finally time to eat. My brothers stretched and rubbed their eyes as we made our way downstairs to the church hall, in the basement. The room was filled with long tables and church sisters wearing white aprons over their Sunday clothes, dishing food onto paper plates: crispy fried chicken, turnip greens, homemade biscuits, and macaroni and cheese. "Thank you Lord," I said, as I shoved a chicken leg into my mouth. "Thank you for all this good-ass food!"

After dinner all I wanted to do was lean back and take a nap. But Mama had other ideas. It was hustling time! She took me by the hand and dragged me over to speak to the pastor, who was standing near the doorway talking to the lady in purple and her husband, who was half her size.

"Look real sad," Mama hissed at me as we walked over. "Pretend you is *lost.*"

When our turn came for time with the pastor, Mama spoke in a voice I'd never heard before, high and tight, like a little girl. "Pastor," she squeaked, "I got five kids I'm taking care of all by myself." When she said this, she pulled me to her chest and gently kissed the top of my head. Her tenderness startled the shit out of me, and I could feel myself go as stiff as a board in her arms. From across the room, my brothers were pointing at Mama and laughing so hard at her bullshit that Andre squirted milk out of his nose.

"We really struggling," Mama continued, without missing a beat. "I sure could use a blessing." The preacher looked down at me and then pulled Mama off to the side so they could talk in private. I watched him take her hands in his and the two of them lean forward in prayer. I don't know what those two talked about, but as soon as we got home from church Mama had an announcement.

"We all getting baptized!" she said.

"Who?" asked Dre, looking up from where he was kneeling by the front door practicing his lock-picking skills.

"All y'all."

"Why?" asked Andre.

"'Cause I said so," explained Mama, cracking open a beer. "We joining the church. Now don't ask me no more questions. Jesus don't like that. Didn't y'all listen to a gotdamn word the preacher was saying?" She raised up her hands as though she was testifying, only she had a cigarette dangling from her lips. "O-B-D-ence!" she hollered. "That's how you niggas is gonna get to heaven."

The next Sunday we got up, got dressed, and walked back across the street to church, where we all got dunked in icy cold water in the name of the Father, the Son, and the Holy Ghost. After we got baptized is when Mama got invited to the pantry. It was everything Miss Cynthia had said it would be.

Sister Ernestine took out a cardboard box and started loading it up with rice, flour, powdered milk, a bag of sugar, Tang drink mix, three cans of sardines, and a jar of Jiffy peanut butter. "Thank you, ma'am," Mama kept saying. "I'm just trying to feed my babies. That's all I'm tryna do . . . Praise Jesus."

"God is good all the time, all the time God is good," Sister Ernestine answered. "Through Him *all* things is possible."

"Yes, ma'am."

"God's work done God's way will never lack supplies," she added. "Chile, I heard that on the TV."

Greater Springfield Baptist Church was the first congregation we joined. But Mama didn't discriminate. She thought we should go see what kinds of charity other churches had to give. Her baptism hustle took us all over town. We became members of Mount Zion Baptist, New Jerusalem Baptist, Bethel Baptist, Shiloh Missionary, Free for All Baptist.

One Sunday Mama even doubled up, taking us to two services in a day. I don't know what the pastor thought when we all showed up to get baptized with our hair already wet.

CHAPTER 4

Angel in Leather Boots

Miss Thompson was my regular third grade teacher, but she didn't teach me much of anything, unless you consider giving the side eye a skill. All my most important learning happened Monday, Wednesday, and Friday afternoons from one o'clock to 2:05. That's when I went downstairs to a small classroom across the hall from the cafeteria to see Miss Troup for Title I remedial reading. Kids called it the slow class, but I didn't care. I loved hanging out with Miss Troup. She was the exact opposite of my mother, quiet, calm, and patient. Plus she was the number one sharpest dressed teacher in the whole school. Miss Troup must have had a closetful of pastel-colored skirt suits and matching floral blouses, because it seemed like she wore a different outfit each day. She styled her hair in a big, bouncy press 'n' curl and wore long fake eyelashes. But best of all were her boots: bad-to-the-bone, knee-length, brown leather with stacked heels, and always polished to a shine. She looked like she just stepped out of the pages of *Jet* magazine. Miss Troup—the baddest bitch at English Avenue Elementary—is the teacher who finally taught me how to read.

"Patricia, honey, just try to sound it out," she said one afternoon, tapping the page with a bright red fingernail painted the color of a cinnamon Red Hot. We were sitting in her classroom with a book cracked open on the desk in front of us. It was hot and I was tired. "Just give it a try," she said again. I looked hard at the letters. I knew they were strung together in words I should recognize, but none of it made sense.

Not everybody in my life knew I couldn't read. Mama, for one, thought I could read my ass off. Every afternoon, she would pick up a copy of the *Atlanta Journal-Constitution* at the corner store. First she'd flip to the comic strips and put her face close to the paper, looking for numbers she was convinced were hidden in the hair or clothes or scenery of the cartoons. Like maybe she'd see a 4 in the background behind Charlie Brown, or a 7 in Hagar the Horrible's beard. Those were the numbers she'd use for the fifty-cent bets she placed with the Numbers Man. The only other thing she got the newspaper for was to check her daily horoscope. But Mama had never finished elementary school and didn't know how to read. Instead, she'd hand me the paper and ask me to tell her what it said.

"Sagittarius," I'd say, looking down with my forehead wrinkled in fake concentration. Then I'd make some shit up: "This is NOT a good day for beating on your children. TODAY IS NOT A GOOD DAY FOR ANY TYPE OF WHOOPING AT ALL!"

Reading Mama's fake horoscope was easy, but with Miss Troup reading took all my brainpower. Sitting beside her, I could feel the frustration rising up inside me. "Sound it out," she said again. "Just take your time." On the page in front of me was a drawing of a cat wearing a big-ass striped hat and red bow tie, holding an umbrella. *What the hell?* I thought. If the damn picture made no kind of sense, how was I supposed to figure out the words?

"I don't know what it says," I mumbled.

"Just give it a try," she urged.

"I can't do it."

I put my face down on the desk, closed my eyes, and swung my legs hard, kicking the metal frame of my chair with the back of my heel.

Bang

Bang

Bang

Bang

I could feel the tears coming. I didn't even know why I was crying. "It's okay," Miss Troup said softly, rubbing my back. "You're doing fine." I thought she was going to let me sit out the whole lesson with my face on the desk, the way Miss Thompson did in my regular class. But instead she told me to sit up. Then she turned her chair to face me, looked me in the eye, and said, "Patricia, I'd like you to come by my room tomorrow morning before the first bell. Do you think you can do that?"

"Why?" I asked, worried. "Am I in trouble?"

"No, not at all," she answered, gently. "I want you to come in early because I have a little something for you."

Then she smiled at me, big and wide, with her cherry-red lipstick and Chiclet teeth, and I got the feeling that whatever she had for me had to be something good.

THE NEXT MORNING I SPLASHED some cold water on my face, pulled on the musty jeans I'd been wearing all week, and took off running—past Mama asleep on the living room sofa with a Bumpy Face bottle on the floor beside her—and out the front door. I flew past Drunk Tony hanging out on the corner. "Girl, you gettin' some ass!" he yelled at me, like always.

"Fuck you, Tony!" I shouted back, and kept on running. When I got to the school, I flung open the side door, ran up the ramp, and busted into Miss Troup's room, sweaty and out of breath.

"Good morning, Patricia!" she said, looking up from her desk. She was wearing a peach-colored dress with a giant bow at the collar and her bad-bitch leather boots. "I'm so happy to see you." Miss Troup reached under her desk, pulled out a blue nylon gym bag, and told me to follow her down the hall to the girls' bathroom. We stepped inside and she started taking things out of her bag and setting them down on the side of the sink: a brand-new bar of Ivory soap, a pink container to put the soap in, a Tussy cream deodorant, a tube of Aquafresh toothpaste, a white washrag folded into a little square, and a brand-new toothbrush with a red handle, still in the wrapper.

"Patricia," she said, turning to me, "these are *your* things."

I looked at the items she'd laid out, then back at her, confused.

"They're for you to wash up," she explained. "I'm going to step outside and give you a little privacy. While I'm gone, I want you to use the washrag and clean your face and neck and under your arms, and put on the deodorant. When you're finished, you can change into these." She reached back into her gym bag and pulled out a bright yellow and white striped T-shirt and a brand-new pair of jeans. I'd secretly been hoping that Miss Troup was going to give me a pair of knee-high leather boots and a big curly wig. But a pair of stiff new jeans from Woolworth and a fresh top were almost as good.

"I'll be right outside," she said, giving me a little pat on the shoulder before she turned to leave, the door swinging closed behind her.

Alone in the girls' room, I let the hot water run over my hands and lathered up the soap. I washed my face and neck and under my arms, just like she told me. I brushed my teeth until my mouth felt minty fresh. Then I pulled on my brand-new clothes and stepped into the hallway where Miss Troup was waiting on me.

"Don't you look right cute!" she exclaimed "Just like a doll,"

I smoothed my hands down the front of my T-shirt. In my whole life I'd never felt as good in an outfit as I did that day. "Thank you," I said, giving Miss Troup a smile so wide I felt like my face was gonna crack in two.

I started washing up at school every morning after that. Sometimes Miss Troup would bring me new clothes I'd never seen before, and sometimes she'd bring back my funky-smelling Goodwill clothes, only they'd be clean and pressed, like she was running a little laundry service at night. She even did my hair, combing out all the naps and braiding four neat plaits that she finished off with plastic barrettes in colors to match my outfit.

One afternoon, Porsha and Mercedes and a bunch of their friends ran up on me and Sweetie as we were walking home. "Look at this nappy-headed ho," said Mercedes, smacking Sweetie in the back of her head so hard it knocked my sister to the ground. "You a nasty bitch! I'ma whoop your ass for being so nasty." There were so many of them, there was nothing I could do but watch helplessly as Mercedes kicked the shit out of Sweetie while the other girls laughed.

Porsha turned to me, taking in my outfit. "Huh," she said, eyeballing the dark blue skirt and crisp white T-shirt Miss Troup had given me that morning. "Patricia don't look so raggedy today. Just her raggedy-ass sister who needs to get beat."

I DON'T THINK MISS TROUP had any idea of the ass-whoopings she saved me from, or that I loved everything about her. I loved the way she smiled and her soapy smell. I even grew to like the *tap tap tap* sound of her shiny red nails hitting the page while she taught me to read. Miss Troup was badass and beautiful. She was like an angel to me.

In the girls' bathroom one morning, I studied her reflection while she fixed my hair. Her hands felt warm against my scalp as she smoothed down my edges with Blue Magic. Whenever Mama did my hair, which wasn't often, she was rough and impatient and it hurt like hell. If I dared cry out from the pain, she'd smack me in the side of my head with her wood-handled brush.

Miss Troup looked up and caught my gaze in the mirror. She flashed me a smile, then her face turned serious. "Patricia," she said, "I want to tell you something."

"Yes, ma'am."

"I want you to know I believe in you."

"Okay."

"You're a bright girl with a lot of potential."

"Potential?"

"It means if you work hard, you can do anything. If you study, and really apply yourself, you can finish school, go to college, and grow up to be anything you want: a teacher, a doctor, a nurse. Anything. You can be *somebody*. Do you understand?"

"Yes, ma'am," I said. But I wasn't so sure. No one in my family did any of the things Miss Troup was talking about. I couldn't name a single relative who'd finished high school. And I sure never saw any of them with a career, or even a legal job for that matter. My uncle was an expert at picking locks, and Aunt Vanessa made money selling food stamps for fifty cents on the dollar. I was pretty sure that's not what Miss Troup was talking about when she said I had potential.

"This world is filled with possibilities," she continued, resting her hands on my shoulders and staring into my eyes. "You can do anything you put your mind to. Anything at all. All you have to do is dream. Promise me you'll remember that, Patricia. Promise me you'll dream."

Nobody had ever talked to me like this before: not Granddaddy, not any of my regular teachers, and definitely not Mama. In my family, we all moved in the same direction, hustling and scheming and getting nowhere. That was the path laid out in front of me. But now here was Miss Troup—in all her leather-boot and red-fingernail finery—telling me I could go another way. I took a deep breath and gazed at my reflection. *You can do anything and be anything,* I thought, trying it on for size. But I wasn't totally convinced I had a place in Miss Troup's world of "possibilities" and "potential."

"Promise me you'll dream," she said again.

"I promise?" I said, looking up at her uncertainly.

"C'mon now, Patricia. I know you can do better than that." She gave me a little squeeze.

"Okay," I said, starting to giggle. "I *promise!*"

CHAPTER 5

Devil in Disguise

Mama had a lot of ideas that made sense only to her. Like the time she decided to cook dinner out in the yard. I was ten years old and we'd moved to a run-down duplex at the bottom of a hill in a shitty part of town known as The Bluff. We didn't have any gas in the house because it got cut off from Mama not paying the bill. So she went out and bought herself a charcoal barbecue grill, which she set up on the screened-in porch, right outside our front door.

The only problem was that grill wasn't made for frying up a skillet full of catfish, like Mama used it for. One evening while she was cooking dinner, the whole porch filled up with thick black smoke. It was so bad that Mr. Willie, who lived in the other half of the duplex, came outside and started hollering.

"Mildred!" he yelled. "Bitch, you tryna kill me?"

"Mind your gotdamn business, you high-yella muthafucka!" Mama yelled back.

They kept up hollering at each other until Mr. Willie decided there was no reasoning with Mama, and called the fire department instead.

The fire truck pulled up to the house with sirens blaring. Mama stepped out of a cloud of black smoke with a fork in her hand, and asked, real casual, "What the hell going on out here?" like her stupid ass wasn't the reason for all the commotion. When the fireman told her she had to move her grill off the porch before she burned the whole place down, Mama threw up her hands in exasperation: "Where I'm supposed to cook then?"

"It's up to you, ma'am," said the fireman with a shrug. "As long as you keep the grill outside." That's when Mama moved her little cooking operation to the front yard. She'd be out there in her faded housedress and a plastic shower cap pulled over her Jheri curl, like she was in the privacy of her own kitchen, not out on full display. As hungry as I was, I would pray for the middle of the month when Mama would run out of food stamps and was low on food, and stop cooking in the yard. Eating ketchup sandwiches for dinner was better than getting teased all day long by kids in class who passed Mama on their way home from school. I felt like I was living in hell. Every day I wished for someone to come along and save me. That's when Mr. John showed up.

SWEETIE AND I WERE SITTING on the porch handclapping "*Miss Mary Mack*" the first time Mr. John offered to buy us food. He pulled up to the curb in front of Mama's house, leaned out the window of his green El Camino, and yelled, "Hey, you girls hungry? I'm finna go to Church's to get something to eat. Y'all want to come along?"

Mr. John was a bricklayer with a big friendly smile. He lived up the street from us, and earlier that summer he'd given my older brother Dre a job helping him mix concrete. In the afternoon, Mr. John would drop Dre back home dog tired and covered in dust. I was in the kitchen with Mama when he came inside one day to say hello.

"How you doing, Mildred?" he asked.

"Oh, you know," she said, pulling on a smoke. "Gettin' by."

Mr. John looked around the kitchen. I saw his eyes move from the burned-out birthday candle stubs stuck to the counter, to the ashtrays overflowing with cigarette butts, to the stove with nothing cooking on it. I guess he figured that "gettin' by" meant "broke as hell," because the next day he showed up with a couple of bags of groceries from the Super Saver filled with bread, spaghetti, grits, pigs' feet, fatback, dried beans, rice, and a package of bologna. He also brought Mama a six-pack of Schlitz Malt Liquor and a dime bag of weed. "You didn't have to do all that," she said, giving him a big smile. "You a good man."

Mr. John started dropping by a lot after that. He'd sit at Mama's kitchen table and laugh at her jokes. The attention put her in a jolly mood, which made things easier on us kids. By "easier," I mean she stopped acting like she was gonna blow our heads off every time she got mad.

Mama always had a quick temper. But she was a tiny woman, barely one hundred pounds, and with all of us kids getting big, it was getting harder for her to whoop our asses like she used to. Instead, when she'd get mad because we left the dishes in the sink, she'd reach into the side pocket of her painter's pants, grab her little .22 pistol, and shoot into the air. "I told you to clean the gotdamn dishes!" *POP POP POP!* Mama fired that pistol the way other parents raise their voice. Every time she got aggravated, we feared for our lives. I used to wonder, *If we so poor, where the hell you getting all these bullets from?* But after Mr. John started coming around, Mama didn't shoot as much. The two of them would talk and laugh and drink their asses off, just like a couple of teenagers in love. Then he'd get up and go back home to his wife.

I don't know what *Mrs.* John thought of the arrangement, but Mama was happy as a pig in shit. Suddenly she had everything that was missing from her life: attention, groceries, and beer. Mr. John would bring her a forty of Schlitz in the morning before he went to work, and a six pack in the evening. Thanks to him, Mama stayed buzzed, and we weren't hungry. It felt like Mr. John coming around was the best thing to happen since we left Granddaddy's liquor house. Then there he was one afternoon, leaning out his car window offering to take me and Sweetie to lunch.

"Y'all hungry?" he asked again.

The two of us ran over and slid into the front seat beside him, giggling about all that crispy chicken we were about to eat. Sweetie sat in the middle, I was by the window. If you'd seen us that afternoon, you would have thought, *Look at that daddy taking his two little girls out for a drive!*

We were a few blocks from the house when Mr. John said, "I got to make a stop real quick." He took a left at the next corner and flipped on the radio. The sound of Marvin Gaye filled the car.

> *Get up, get up, get up, get up*
> *Let's make love tonight*
> *Wake up, wake up, wake up*
> *'Cause you do it right*

Mr. John sang along as he drove us up the street, past Booker T. Washington High School. He had a nice voice, I remember. Deep, like Barry White.

He turned left again, onto Ralph David Abernathy Boulevard, toward the low stone entrance of the Westview Cemetery. Then, without saying a word, he drove his car through the gates.

I figured Mr. John had some dead relative he needed to see. Why else would we be driving through the grounds, past the rows of tombstones that seemed to go on forever? I was about to ask him, "Who died?" when he turned off the main road and pulled to a stop under the shade of a giant willow tree.

"Why we here?" I asked.

He turned to me and smiled. "We gonna play a game."

"What kind of game?"

"A singing game," he said. "Rabbit, turn your head and look out the window. You gon' sing until I tell you to stop. You do a good job and I'll give you five dollars."

"What I'm supposed to sing?"

"It don't matter."

"I don't know any songs."

"You know that song that was just on the radio, don't you?"

"I guess."

"So sing that."

I didn't understand this stupid-ass game. But I did understand "five dollars." So I turned my head and stared out the window. Not far from where we were parked, somebody had left a bunch of flowers tied with a white ribbon on a grave marked by a stone that said in big block letters, MOTHER.

"Sing, Rabbit," said Mr. John.

"Get up get up get up let's make love tonight," I began. Behind me I heard the sound of the driver's-side door opening, the rustle of leaves, and Mr. John whispering to Sweetie, "I ain't gonna hurt you."

It was hot as hell in that car. I could feel beads of sweat dripping down my back. I didn't know the lyrics. "Ooooooh baaaaby," I sang, off key. "La la la motion, like the ocean, magic potion baaaaaby . . . *get up get up get up.*" I rested my head against the window and closed my eyes.

It felt like a long time had passed before Sweetie finally walked around the front of the car and pulled open the passenger-side door. I slid over to let her in. Her mouth was shut tight, her head down. She wouldn't look at me. Instead, she stared at the floor, twisting her fingers into knots in her lap. Mr. John said, "Okay, Sweetie. You look out the window now." Then he turned to me.

"Lay back," he said.

"Why?"

Suddenly his hands were on my shoulders, pushing me down. "Hey . . ." I said, struggling to sit up. But he was so strong, pulling my legs toward him, until they were dangling out the car door.

I could feel his hands tugging on my shorts. "No!" I cried, gripping my waistband and holding on. "Noooo!"

Mr. John leaned over me with his face near my ear. "Let go, Rabbit," he said, real low. "I ain't gonna hurt you. It's gonna feel nice, like kissing. That's all I'm gonna do."

My heart was racing and I felt trapped, pinned down by Mr. John's words and his big hands all over me. He knelt in the grass and pulled on my shorts. I felt my bare bottom against the car's leather seat, then his wet mouth between my legs. I grabbed at the hem of my T-shirt, trying to pull it down to cover my privates.

I looked up through the windshield at the leaves against the blue sky high above.

I pretended to fly away.

AFTERWARD, MR. JOHN TOOK US to Church's just like he promised. And gave us each a crumpled-up five-dollar bill. "You know I help your mama out so y'all don't go hungry," he said, pulling up to the curb in front of our house. "If y'all tell her about this, she'll be mad. She'll whoop your ass *real* good. I won't be able to come over and help out. Then y'all be hungry for real." He told us this every time he took us to the cemetery, which he did for years.

Sweetie and I never talked about what Mr. John did to us in the front seat of his El Camino, but I always wondered if Mama noticed what was going on. I wondered if she ever asked herself where her boyfriend took her babies, or what he was paying for with those five-dollar bills. I know Mr. John gave Mama attention and he kept us all fed. Maybe when she looked at him that's all she wanted to see.

CHAPTER 6

First Dance

Puberty hit me like a brick to the chest. By the time I was twelve, my face was a mess of acne, and my dried-out Jheri curl looked like somebody set a bowl of burned curly fries on top of my head. To make things worse, I was suddenly busting out all over the place with curves that made my T-shirts stretch tight across my chest. Compared to me, Sweetie had it easy. She had good hair, clear skin, and dimples. She wasn't just pretty, she was *grown*. She wore eyeliner and lip gloss, and when she walked she switched her hips in a way that had grown men following her down the street. "*Daaaamn,* girl!" they'd say. She was only two years older than me, but next to her I was invisible.

As jealous as I was of her good looks, I couldn't get away from Sweetie even if I wanted to. Mama had moved us again, this time to a one-bedroom duplex on Baldwin Street, and Sweetie and I shared a small room right off the kitchen. By "right off the kitchen," I mean we were actually *in* the kitchen. At night, Mama laid a mattress on the floor and hung a thin sheet up to separate our room from the fridge and the stove. If I woke up sweaty in the middle of the night, I could roll over, open the fridge door and stick my head inside to catch a breeze. With Mr. John still messing with me, my personal AC was pretty much the only good thing I had going on.

Back then, Sweetie couldn't be bothered with me. She was best friend with our cousin Peaches. On Saturday nights, the two of them would smoke Newports, drink forties, and head to the teen dance at the YMCA rec center. Then they'd spend the rest of the week whispering and laughing about all the good times they were having without me. It drove me crazy. All I wanted was for them to let me tag along so I could get out of Mama's house and have me some fun, too.

"C'mon," I begged Sweetie one night when the two of them were getting ready to leave, filling our side of the kitchen with the smell of cheap Primo perfume they stole from Woolworths. "Why can't I go with you?"

I expected her to say the same thing she did every other time I'd asked: "Nah, you too young," like we weren't practically the same age. But it turned out to be my lucky night. Sweetie's boyfriend, Crispy, was bringing along a dude named Fresh, and he didn't have a girl.

"You can hang out with him," said Sweetie. "He prolly ugly as fuck anyway. So you two be a perfect match." Peaches and Sweetie bust out laughing like a couple of hyenas. But I didn't care, I was gonna get my dance on. I left the kitchen and went into the bathroom to practice my moves in front of the cracked vanity mirror. I could only see myself from the chest up, so I focused on getting the top half of my body right, snaking my neck from side to side, and hoped that the bottom half of my body was following along.

When the boys came by to pick us up later that night, I could see Sweetie wasn't lying; Fresh was corny as hell. He was rocking a long-in-the-back Lionel Richie Afro and British Knights tennis shoes laced up real tight. "Wut up?" he said, giving me a shy nod.

"Nigga, why your shoes so tight?" I asked, with my hands on my hips. "Your mama tie them like that so they don't fall off your feet? You look like you take the short bus to school, like you *re-tarded.*" Sweetie, Peaches, and their boyfriends, Crispy and Mike, fell out laughing.

"Yo, your sister can jone," said Crispy. I could tell he was impressed. Joneing—or what old folks called "the dozens" and white people called "being mean"—was a survival skill I'd recently picked up. I learned that if I could crack a joke—about somebody's musty clothes, nappy hair, stanky breath, or pretty much anything about their mama—before they had a chance to jone on me, folks would leave me alone. I didn't know all that when Mercedes, Porsha, and them were coming for me. But by the time I hit sixth grade I'd found my secret talent: I had a lot of mouth.

Fresh didn't talk much on the way to the dance. He just shuffled beside me looking down at his sneakers, while the rest of them walked ahead smoking Newports and trying to act cool. Maybe everything would have been different if I'd stuck with Fresh and his corny ass. He wasn't a greasy Jheri-curled bad boy like the type my sister went for. He barely had the nerve to look me in the eye; he was harmless. I would have been okay with a boy like that. Instead I met Derrick.

We were a few blocks away from the dance when I first heard the music.

> Freaks come out at night
> Freaks come out at night

Somebody was blasting Whodini through the open window of a beat-up Chevy Nova, and they were headed our way. "Yo!" the driver yelled. "Yo, Crispy!" He made a U-turn, pulled up beside us, and stepped out of his car.

Crispy gave his friend a pound then turned to us. "This my boy, Derrick," he said.

Derrick was short, with a chipped front tooth, and wearing tight-ass jeans with stiff creases down the middle of each leg.

"Where y'all headed?" Derrick asked.

"To the dance," said Mike.

"Oh yeah?" Derrick leaned back on his car.

I watched his eyes move from Crispy to Sweetie to Peaches, then land on me. He eyeballed me like he was dead broke and I was a pay-what-you-can hooker. He cocked his head to the side and stared harder.

"What you looking at?" I asked.

"At you!" he said. "You got a big ole butt."

"Fuck you, you dumb-ass," I shot back. "Your pants is so stiff if you bend down you gonna break both your legs. Robot-looking muthafucka."

For a second nobody said anything. Then Derrick bust out laughing: "Damn! Girl, you *crazy!*"

"Yeah," said Fresh. "Her ass crazy as hell."

I guess "crazy" was Derrick's type, because the next thing I knew he was telling us all to get in his car so he could drive us to the dance. "You sit up front," he said, grabbing my arm. "You gonna ride with me."

THE DANCE WAS ON FIRE. It was the summer of 1984, the early days of hip-hop, and the rec center was filled with kids locking, popping, and uprocking. "Candy Girl" came on and Peaches, Sweetie, and all their friends busted out the Cabbage Patch, which they'd been practicing to perfection. I did my signature move, the snake, rolling my body from side to side. But it was Derrick who stole the show. That boy could moonwalk just like Michael and pop the splits better than James Brown. I was checking him out from the corner of my eye when "Purple Rain" came on and suddenly he was grabbing me from behind and humping my ass like a dog in heat.

"Get the fuck off me," I yelled, peeling his hands off my butt. He just laughed and humped me some more.

"Purple raaaaaaaain," he sang in my ear, pulling me close.

At the end of the night, Derrick told everybody he was going to drive me home. "She not gonna do nothing," Sweetie said, leaning her head into the driver's-side window. "She's a *virgin*." She laughed, grabbed Crispy's hand, and wrapped his arm over her shoulder. You could tell by the way the two of them were walking down the street, hugged up like they were glued at the hips, that they were on their way to do some very non-virgin activities.

"That true?" Derrick asked, turning to me. "You really a virgin?"

"Yeah."

"Cool," he said, pulling away from the curb. "I like that."

I was proud of my status. In my mind, there were only two ways for a girl to be, a virgin or a ho. Thanks to Mama's baptism hustle, I had been all over town hearing preachers hollering about the evils of fornication. I knew God didn't want me to be no ho.

A few minutes later, we pulled up in front of our duplex. I opened the door to get out, but Derrick put his hand on my knee. "Hold up," he said. "I wanna talk to you."

"'Bout what?"

"'Bout you."

He wanted to know what kind of music I liked, what movies I'd seen, and who I lived with. "You smoke reefer?" he asked.

"Nah."

"Drink beer?"

"I don't do none of that."

"What's your favorite movie?"

"*Breakin'*."

"Yeah, that movie's fresh to death," he said. "You ever been roller-skating?"

"Nah."

"I should take you some time. You cute. I like you."

For a minute I thought I was hearing things. But then he said it again.

"I like you. I'ma take you *skating*."

No one in my whole life had ever told me I was cute. And no one had ever said they wanted to take me anywhere, except for Mr. John, but that didn't count. I sat in Derrick's car and stared at my hands, trying to stop my heart from beating so loud.

"How old are you?" I asked. I knew Derrick had to be older than me, since he had a car. But I wasn't expecting the answer he gave. "Twenty," he answered. He turned to look at me. "How old are *you*?"

Derrick had eight years on me, but I didn't want him to think I was a little girl. So I lied. "I'm eighteen."

He raised his eyebrows. But he didn't say a thing.

WE STAYED IN HIS CAR TALKING until the sun came up. It was five in the morning when I finally went inside and took my ass to bed. I was so tired I didn't even remember to put a plastic shower cap on to keep my Jheri curl juice on my hair, where it belonged. When I woke up to the sound of Mama calling my name, there was curl activator grease all over my face.

"Rabbit!" she yelled from the other side of the bedsheet hanging in the kitchen. "Some pea-headed boy here for you."

"What?" I said, rubbing my eyes.

"Some boy at the door say he looking for you."

I rolled off my mattress, pulled on my jeans shorts, and stepped into the kitchen. Mama was at the counter, cutting up a turkey wing with her big kitchen knife. "Don't just stand there," she said, looking at me. "Go see who it is."

The only boy I could think of who might come to see me was Petey, my friend who lived up the street. Sometimes the two of us would hang out together and talk about football and practice kissing. Petey was two years older than me, in ninth grade. He was short and thick and had big dreams of being a running back for the Georgia Bulldogs. We'd talk football and I'd watch him "pump iron," which consisted of him doing curls with the one twenty-pound barbell he and his six brothers shared between them. One time when I was at his house, after he finished describing to me the "dope-ass" play he ran in JV practice, he pushed me down in his bed and tried to grab on my titties. I was so mad, I pinned him in a figure four and punched him in the throat. "We just friends, nigga!" I yelled in his face.

But it wasn't Petey at the door. It was Derrick, dressed in the same clothes he'd been wearing the night before, only the crease in his jeans wasn't as sharp.

"Hey," I said, letting him into the living room. "What you doing here?"

"I came by to see if you had a good time last night."

"Why?"

"Because I like you."

I looked at the floor, feeling my face getting hot. Mama must have been listening from the kitchen, because she came back into the living room holding her cooking knife.

"Hey," she said, squinting at Derrick. "You like my baby?"

"Yes, ma'am," he said. "We friends."

"She tell you how old she is?"

"Yeah, she say she eighteen."

Mama let out a snort. "Girl, stop lyin'," she said, heading back to the kitchen. "You know your hot ass only twelve. You hear that, boy? She *twelve*."

Derrick shot me a glance. But he didn't get up to leave. He didn't do anything except tell me my body didn't look like no twelve-year-old's. He reached out and gently pulled me to him, rubbing on my arm. "You got a pretty smile," he said, real low so Mama couldn't hear. "And nice lips," he whispered.

I didn't know what was happening in my body. It felt like somebody had dumped a cup of baby mice in my belly and they were running around, making me all tingly. Derrick was making me feel good. I wanted that feeling to last forever.

On top of the TV was a pencil and an empty pack of Mama's

Winstons. I ripped open the cigarette box, spread it flat on top of the set, and slowly, in my best penmanship, wrote Derrick a note: "Will you be my boyfriend?" Underneath I made two boxes, one marked YES and the other NO.

Derrick gave me a funny look when I handed him the paper.

"Rabbit . . ." he started to say.

Then he stopped himself.

He picked up the pencil and put a check in the box marked YES.

CHAPTER 7

Love Lesson

Before Mr. John started coming around, messing with me and my sister and buying Mama food, my mother actually had herself a real good man. Mama's boyfriend Curtis first lived with us at the liquor house. On Sunday mornings while Mama was sleeping off her drunk, he would take me and Sweetie out to Grandma's Biscuits for breakfast and let us order whatever we wanted. And when Mama was beating us too crazy with her leather belt, he was the one who would step in and tell her to take it easy. "C'mon now, Mildred," he'd say. "They just being kids."

Curtis worked out behind the Grey and White Auto Parts, fixing cars in the parking lot, under the shade of a big oak tree out behind the shop. He stood five foot four—his friends called him Shorty—and he had a receding hairline. If you squinted, he kinda looked like George Jefferson, if Mr. Jefferson dressed in grease-stained jeans and work boots instead of three-piece suits. Curtis wasn't much to look at, but everybody at the liquor house knew he treated Mama right. Aunt Vanessa used to say, "You better hang on to that one, Mildred. You got a good one right there," which meant Mama had a man willing to take care of a bunch of kids who weren't even his.

After Granddaddy got locked up for shooting Miss Betty, Curtis moved Mama and all us kids into a little house across town, on Oliver Street, with three bedrooms and a yard out back where Mama planted vegetables and kept a little chicken coop. With Curtis, Mama didn't have to worry about a thing. He paid the rent and took care of all the bills. He even bought Mama her very own car, a pale pink '69 Cadillac Coupe de Ville. Mama called her ride the Pink Panther and drove it all over town playing B. B. King on the eight-track and singing at the top of her lungs, *Rock me, baby, rock me all night long!* But of all the things Curtis did for Mama, the best he ever did was getting her some brand new teeth

For as long as I'd known Mama, she never had any front teeth. Aunt Vanessa told me that when I was a little baby, my daddy had balled up his fists and knocked Mama's teeth right out her mouth. "That's when your uncle chased him off with a shotgun," Aunt Vanessa said. Mama told me my father was a "no-good downtown clown" who never did anything for her except beat her ass and treat her like dirt.

Curtis was Mama's chance to do things right.

THE DAY CURTIS BROUGHT MAMA HOME from the dentist, all us kids gathered in the kitchen to stare at Mama's mouth. She rested her elbows on the table in front of her and put her hands up to her face, fanning out her fingers and smiling wide.

"Ooooooweeeee!" said Jeffro. "Mama got her front door fixed."

"How I look?" she asked, excited. Then she answered her own question. "I look *good*!"

Actually, she looked crazy. Those dentures were too big for her face. Mama looked like she was wearing a set of donkey teeth. But I sure as hell wasn't going to be the one to tell her she got the wrong size teeth. "Took Curtis six months to pay off the dentist!" she exclaimed. *"Six months!"*

She turned to Curtis: "You got a sister looking just like Diana Ross. I'm finna go down to the Grey and White and smile at all the mens!" She let out a giant laugh. That's when her donkey teeth slid out of her mouth and fell on the table. "Oh!" she said, scooping them up and sticking them back in. "Guess I just gotta get used to them."

Too bad for Mama, those dentures weren't as easy to get used to as she thought. She'd wear them for a few hours and then they'd start to hurt, so she'd take them out to rest her mouth. I'd find them on the kitchen counter, or on top of the TV. Once, after they'd been missing for days, I found them in the back of the freezer. Still, Mama loved those teeth. She said they made her feel as sharp as Katherine Chancellor on *The Young and the Restless*. I didn't see the similarity, unless you count the fact that the two of them were both drunks. Katherine Chancellor sipped her liquor out of crystal glasses; Mama sucked her gin straight from the bottle.

As much as Mama liked her new look, something changed not long after she got her new teeth. It was like a switch flipped, and Mama got it in her head that the only reason Curtis was being nice to her was because he was up to no good. Suddenly she was angry all the time.

Curtis would come home from work, and she would meet him at the door with her hand on her hip and start interrogating him like she already knew he was guilty: "Where you been all day, Curtis? You think I don't know what you up to?"

He'd look at her, confused. "I been at work, and you know that."

But Mama wouldn't let it go. The drunker she was, the worse it got.

Curtis came home one day looking dog tired and stinking of sweat and motor oil. Mama followed him into the bedroom, a half-empty pint of gin in one hand and a Winston hanging from her mouth.

"Where you been, you short-ass grease monkey?" she demanded. I noticed she was gripping the doorframe to steady herself. That was never a good sign. "I said, *Where you muthafuckin' been?*"

Curtis sat down on the edge of the bed to pull off his work boots. He didn't even look up. "Mildred, you know where I been," he said wearily.

"Oh, I know where you been all right!" She pointed a finger at his chest. "Out fucking some hos!"

"No, Mildred. The only place I been is at the Grey and White. I put in a transmission, took me all gotdamn day."

"You a muthafuckin' lie!" Mama shrieked. "You ain't nothin' but a low-down sawed-off little nigga!"

It was hard not to feel sorry for Curtis. Even though he'd spent eight years in the military and could probably kill a man if he had to, he was the quiet type. Most times he didn't even make a sound when he laughed. The only way you could tell he thought something was funny was by his shoulders shaking up and down. It was like his number one mission in life was not to get noticed. But Mama wouldn't get out of his face.

"You gonna hit me, Curtis?" Mama hollered. "That what you gonna do? Try me, nigga! Just *try me!*"

Mama's veins were popping out of her neck. Curtis looked up at her and let out a loud sigh. It was the kind of sigh that said, *How the hell did I end up with this crazy-ass bitch?* Then he lay back on the bed and pulled a pillow over his face. That's what really set her off.

"Oh, hell nah!" she screamed. "Hell to the muthafuckin' nah!"

She stomped into the kitchen, grabbed her dentures off the counter, and stormed out of the house. "You ain't gonna play me for a fool, you Gary Coleman little-dick-having somabitch!" she yelled. "Think you know me? I'ma show you." She kept up her hollering as the screen door slammed behind her: "I'MA SHOW YOU!!!"

Dre, Sweetie, and I ran to the front window to see where she was headed. Curtis was right behind us. Outside we could see Mama, dressed in her faded pink housecoat and no shoes, marching through the yard toward the Pink Panther. We all watched as she bent down and put her dentures on the road, right by the front tire of her car.

"Jesus!" yelled Curtis as he ran out of the house after her. "Mildred, you done lost your damn mind!"

Mama looked over at Curtis running toward her and stepped into the car. She revved the engine just as he reached out to grab the door handle. But he was too late. The Pink Panther lurched forward and rolled right over those fake teeth. Mama leaned out her window and screamed, "Fuck you!" before she backed up and drove over her teeth again.

"*Daaaaaaaamn!*" said Dre, as we watched from the living room window. "Mama's cold-blooded." He sounded impressed. But I couldn't take my eyes off Curtis.

Outside on the curb, he put his hands on top of his head like he was trying to keep it from exploding off his body. Then he turned his back on Mama, and walked slowly toward the house.

He left us after that. He said, "I can't take this shit no more," packed up his work boots, his oil-stained jeans, all his tools, and moved out. Mama must have known it was her own crazy ass that ran him off, but she still seemed surprised to see him go. "Why?" she wailed, as he walked out the door. "How you gonna leave me with all these damn kids?" Nothing she said made a difference. Curtis was done.

THINGS GOT REAL BAD AFTER THAT. Mama went from being a regular alcoholic to one of those drunks who didn't do anything but cry. She'd cry and drink her gin and cry and whoop us kids. She'd sit at the kitchen table all night long howling about how nobody loved her. Sometimes Mama got so in her feelings about Curtis being gone that she'd put me and Sweetie in the back of the

Pink Panther, drive over to the Grey and White, and raise Cain in front of all his friends.

"You got company, Shorty," his boys would say when they saw her coming. They'd lean back on their cars, pull their hats down low, and watch the show. Mama would yell and scream like a banshee, demanding Curtis bring his midget ass home. One time she knocked out his car windows with a tire iron; another time she tried to slash his tires with a cooking knife. "I'ma kill you, you muthafucka!" she'd yell. "They gon' bury your tiny ass in a baby casket."

Every time she showed up at the Grey and White, Curtis would hold up his hands and try to calm her down. "C'mon, Mildred," he'd say. "You don't need to be acting like this. Simmer down, now." He'd promised to come by the house later. He'd tell her they'd work it out. Then she'd sit at the kitchen table waiting on him all night long, but he'd never show up. "Don't nobody love me," I'd hear her sob. "Nobody at all."

I don't know what made Mama act so crazy, or how love and anger got so mixed up in her head. All I know is by the time I met Derrick, when I was twelve years old, everything I knew about relationships was what I'd learned from her.

CHAPTER 8

Age of Consent

Sweetie stood on the porch steps with her hands on her hips and a Newport dangling from her mouth. "So," she said, looking down at me sitting on the step beside her. "You givin' that nigga some pussy?"

Derrick and I had been dating three months and this was Sweetie's way of asking me how things were going. I leaned my head against the railing and ignored her. Sweetie and I were sisters, not friends. I wasn't trying to tell her my business.

I had more important things to think about, like the outfit I was wearing. Derrick was on his way to pick me up to take me to Jellybean skating rink and I needed to look fresh. I had on my pink oversized T-shirt that hung off one shoulder *Flashdance* style and some dark blue Jordache jeans that my brother Dre stole off somebody's clothesline. Sweetie said I could have the jeans because they didn't fit her in that skintight way she liked. I took my banana clip out of my back pocket, gathered up my hair in a side pony, and tried to slick down my edges with my hands.

"Gotta give him that cooch," insisted Peaches, who was sitting beside me on the porch step. She had a giant tube of strawberry Lip Smacker she kept rubbing over her lips, like she was trying to get them extra glossy. But she wasn't fooling me. I knew for a fact that she would lick the shit out of that Lip Smacker when she was hungry. "He ain't gonna stay with you if you don't give him some pussy."

The two of them were talking at me like I cared what they thought. They didn't know I didn't give a shit about their opinions anymore. I had a boyfriend, and he was the only one who mattered.

Derrick and I went everywhere together. In the mornings he would pick me up in his Chevy Nova and drive me to Dean Rusk Elementary, where I was the only girl in seventh grade whose boyfriend had a car. In the afternoons I'd walk over to Fish Supreme, where Derrick had a job as a fish fryer, and sit at a little plastic table and wait on him to be done. On the weekends we'd go to Jellybean. I loved how he'd take me by the hand and introduce me to his boys: "This is my lady." Or, "This my girl."

The first few times we went to the rink, he tried to teach me how to skate. But mostly, I just stood on the side and watched him whiz by. He moved like those wheels were part of his body, like he was flying through space.

On the porch, Sweetie exhaled her cigarette smoke. "So did you give it up?" she asked again. "'Cause Peaches is right, he ain't gonna stay with you if you don't give up the bootie." I rolled my eyes. Sweetie didn't know shit about my life.

Sometimes Derrick and I would just sit in his car and talk—*just talk!*—for hours. He'd ask me about school and how my day was, and we'd talk about Mama. I told him how she called me all kinds of names, like bitch and ho, and said I was ugly. He put his arm around me and cooed, "Don't listen to her. She don't know what she's talking about. You look good to me."

IT HAD BEEN A LONG TIME since I'd had somebody to talk to. The last best friend I had was KooKoo, a scrawny little chicken from when we lived with Curtis on Oliver Street and Mama kept a coop in the backyard. After school I'd run home and crawl into that chicken coop and pet KooKoo's little chicken body. After Curtis left, I'd cry to KooKoo about how much I missed him and how much I hated Mama for running Curtis off. That scrawny-ass bird would look at me with her beady eyes like she understood every word I was saying. I'd leave her coop stinking like chicken shit, with feathers in my hair. But I didn't mind. KooKoo was the only one who really cared about me. Then one day I came home from school and ran up the stairs, through the house, and into the backyard. Mama was standing there holding KooKoo by the neck, only KooKoo didn't have a head. My mother had murdered my best friend.

"Rabbit," Mama said, as I held my breath and tried not to cry, "Come down here and pluck these damn feathers." That night, Mama fried KooKoo for dinner and served her up with hot sauce. After that I didn't have anyone to talk to until Derrick came along.

"Damn, your mama be trippin'," he said when I told him how she'd fire her .22 in the house whenever she got mad. We were sitting in his car parked at the curb, like we did almost every night. "She's crazier than an outhouse fly," he added. "But don't worry, I'll take care of you."

Derrick didn't want me for sex the way Sweetie's boyfriend did her. He just wanted to talk. Of course, sometimes we kissed and he'd try to feel on my titties. And he did make me rub his wiener, whispering, "just kiss it," as he pushed my head down to his crotch. But I laughed and said, "Stop, Derrick! You *stoopid*." And he stopped.

I didn't tell Sweetie about Derrick because I knew she wouldn't understand. It was obvious she hadn't been paying attention the way I had when all those different preachers told us, "Sex is for the marriage bed." I was gonna wait for marriage, like we were supposed to. But Sweetie wasn't trying to live right. One time I caught her showing Peaches how to suck a dick with a cherry Blow Pop. There was no way fourteen-year-old out-of-wedlock dick-sucking practice was okay in the eyes of our Lord and Savior Jesus Christ.

I didn't want to talk about sex with my fast-ass sister, but she wouldn't leave it alone. "Diiiiiid yooooooo fuuuuuuck Derrick?" she asked again, this time talking extra slow and flapping her hands in the air like she was doing sign language.

"No!" I finally yelled. "And mind your damn business."

"Urghhh!" Peaches groaned loudly, throwing up her arms like she'd just seen a fumble at the touchdown line. "Stop being such a baby! You think he wants to sit and *talk* all day long? Don't no nigga need you for that!"

"What niggas like," said Sweetie, slowly swiveling her hips like she was dancing, "is a wife in the streets and a freak in the sheets. You gotta give him some of that ass and you gotta give it to him *gooooood*." Then she leaned forward, lifted one leg, and slowly turned in a circle, humping the air like a dog in heat, which I guess was supposed to be a demonstration.

A FEW NIGHTS LATER, the three of us were hanging out with our boyfriends in the kiddie playground at Washington Park. It was late—way past the time kids would be there—so we had the playground all to ourselves. Sweetie and Peaches were sitting on the bottom of a slide, sipping on forties, while their boyfriends were showing off, doing pull-ups on the monkey bars. Derrick and I were on the swings nearby.

I was trying to decide if Derrick would think I was a baby if I asked him to push me, when I noticed Sweetie get up and walk over to Crispy. She pulled him toward her by the belt loops on his jeans, and whispered something in his ear. He grinned and put his arm around her. I watched as they started walking away together, heading out of the playground toward some trees.

"Where y'all going?" I called after them. I couldn't believe Sweetie was leaving. It was her idea to come here in the first place. We'd all been sitting on Mama's porch when she looked over at

Crispy and said, "You wanna go to the park?" She was sucking on a Blow Pop when she said it, only she wasn't sucking on it like a regular person, she was rolling her tongue around in circles, reminding me of how Uncle Stanley's mouth would move after he had a seizure.

"Yo," said Crispy. "The way you sucking that thing though . . . yo." Which I guess meant, "Yeah, the park sounds nice!" because the next thing I knew we were all piling into Derrick's Chevy and heading to the playground.

"Where y'all going?" I called again.

"We out this bitch," Crispy answered, not even bothering to turn around. He threw up a peace sign, then dropped his arm back around Sweetie's shoulder.

"Peaches!" I yelled to my cousin. She was holding Mike's hand and walking in the other direction.

"Bye, girl!" was all she said.

Then it was just me and Derrick sitting on the swings in the cool air. "Fuck those bitches," I said. "Let's go back to Mama's."

"Nah," he said, taking me by the hand and pulling me off the swings. "I got a better idea."

"Where we going?"

"Rabbit, anybody ever tell you that you ask too many questions?" he said. "Don't worry about where we going. I *got* you."

Derrick led me across the playground and toward a patch of trees. The farther we walked, the darker it got, until it was just the stars and moon that lit our way. Derrick stopped, slid off his fake black leather knock-off Members Only jacket, and laid it on the ground at the base of a big oak tree with the inside of the jacket facing up.

"C'mon and sit down," he said, waving his hand like he was offering me a seat on his sofa. "I don't bite."

I sat down with my legs crossed. He knelt beside me and started kissing on my neck. "I really like you, Rabbit," he said. "You my baby. I think I'm falling in love."

SWEETIE AND PEACHES TALKED at me like they knew everything, like they were some kind of experts in sex. But what they
didn't tell me is that once you open your legs for a man, he can't
ever get enough.

"One time" was all Derrick asked for, whispering in my ear as
we lay out under that oak tree. "One time," he said. "One taste, just
a li'l bit. *C'mon, baby, lemme just put the head in. I promise it won't
hurt.*" But I might as well have given Derrick my body wrapped in
a bow, because after that night in Washington Park it felt like he
owned it. He wanted it in the park, in his car, on his sister's living
room floor while she and her kids were asleep.

One night he broke into an upstairs apartment, where a lady
named Catfish used to live before she moved to the projects. She had
bugged-out eyes and a mess of kids, and her apartment had been va-
cant for months, ever since they'd left. Derrick pushed open an unlocked window, climbed inside, opened the front door, and let me in.

He'd brought along an old towel, which he spread on the dusty
floor. "See," he said. "I made it nice for you."

I looked from him to the towel but didn't make a move. There
was no way in hell I was gonna lay down in all that dirt. But before
I could open my mouth to tell him I was leaving, he grabbed my
hand and said, "I love you."

I don't know why those words made me so weak, but every
time Derrick said them I couldn't resist. Maybe it's because no one
in my entire life had ever told me they loved me before. But Derrick
was a pro, feeding me so much sweet-sounding bullshit I felt like I
owed him something in return.

Before Derrick came along, I had big plans to wait until marriage. But with Derrick on top of me and Sweetie's ho advice ringing in my ears, me and my plans didn't stand a chance.

It had been early fall when Sweetie and Peaches stood on
Mama's porch and schooled me on sex. By that winter, both of
them were pregnant. A few months later I turned thirteen. Then
I got pregnant, too.

CHAPTER 9

Love and Options

N o way," said Mama. "I don't believe in killing no babies."

"But Patricia is so young . . ."

"I already told you, no."

Miss Munroe, our caseworker from the Fulton County Department of Family and Child Services was standing in the middle of the living room, with her clipboard in hand, trying to get Mama to listen. But Mama wasn't having it. She just sat on the edge of the ratty brown sofa with her neck muscles popping out like thick cords of rope, and kept saying no.

"No," she said again. "I ain't doing it."

Miss Munroe cleared her throat: "But Miss Williams, if we could just discuss Patricia's *condition* . . ."

"I said, no."

Miss Munroe was short and plump, with a little-girl face and freckles running across her nose. Watching her and Mama, it was obvious she didn't know the first thing about how to fight. She wasn't yelling or cursing or promising to cut a bitch. Her titties were so little she probably couldn't even hide a thin single-edge razor blade in her bra, much less a big wooden-handled kitchen knife like I saw one of Granddaddy's customers whip out during a fight. Instead Miss Munroe just stood there stiff as a board, trying to get Mama to do what she wanted by glaring at her and clearing her throat.

THE FIRST TIME MISS MUNROE came to check on us was when we were living on Oliver Street, not long after Curtis left. I guess somebody—a teacher, maybe a neighbor—reported us kids to DFACS for being so dirty that it looked like nobody was taking care of us. After that, every place we moved Miss Munroe would magically show up in her white Buick, knocking on the front door and calling out to Mama, "Hello, Miss Williams!" in a singsong voice like the two of them were friends. Even back then I knew that was a situation that would never happen in a million fucking years. Mama didn't mix and mingle with white folks and Miss Munroe didn't look the type to have a friend with no front teeth.

The first year Miss Munroe started seeing us was right before Christmas. And the very first thing she did was sign us up for the Empty Stocking Fund, a charity that gives poor kids Christmas presents for free. Before Miss Munroe came into our lives, Mama's idea of celebrating Baby Jesus's birthday was throwing James Brown singing "Santa Claus Go Straight to the Ghetto" on her little record player. But thanks to Miss Munroe, we started getting actual toys, which Mama handed us on Christmas Day in crumpled brown paper bags from the Super Saver.

When I was eight, I got a plastic tea set. The next year I got a baby doll with eyes that opened and closed. The Christmas before I met Derrick, when I was eleven, I got a red and white book bag that said COCA-COLA on it and a knockoff Barbie doll in a pink princess dress. I tried to make Fake Barbie ride on top of one of Mama's empty Schlitz Malt Liquor cans like it was a horse. But the can was too big and one of Fake Barbie's legs popped off and flew right across the room.

It wasn't just Christmas presents that Miss Munroe got us, either. She had all kinds of hookups. She told Mama about Free Vaccinations at the Health Department; Free After-School at the YMCA; and the Free Dentist who came around every summer in a big white truck that made me think he probably drummed up extra business by running a side hustle as the ice cream man. Miss Munroe gave Mama vouchers for Free School Shoes at the Buster Brown store and Free Winter Coats at the Kmart. And now she was trying to give Mama Free Advice about what to do about my pregnant ass. I stood in the corner of the living room and listened to them talk about me like I wasn't even there.

"We should think about the impact this pregnancy will have on Patricia's education," continued Miss Munroe, tapping her clipboard. "There are options you may not have considered . . ."

"What kind of options you talkin' about?" asked Mama.

"Well, adoption is one possibility. There are so many families willing to provide a loving home . . ."

Wait . . . what did she just say? It sounded like Miss Munroe was trying to get Mama to give my baby away. But that couldn't be right. The baby was *mine*. It was already inside me. Giving my baby to somebody else would be like letting me have a doll for Christmas, then snatching it away.

Mama didn't like what she was hearing, either, because she looked Miss Munroe dead in the eye and told her again, "No."

I NEVER UNDERSTOOD THE RELATIONSHIP between Mama and our caseworker. Miss Munroe looked like she was genuinely trying to make Mama's life easier with her vouchers for Free Shit. But every time she came by, she always pulled me to the side for a private conversation, asking me a million questions—"You doing okay,

Patricia? You getting enough to eat?"—like her real mission was to get me to snitch on my own mother.

Maybe that's why Mama couldn't stand her. "Stale-ass cracker," she'd say the minute Miss Munroe was gone. "Coming in here tryna tell me how to raise my own gotdamn kids. That bitch can go straight to hell."

If Miss Munroe knew how much Mama hated her, she never let on. She just kept showing up at our door with a handful of vouchers and a bunch of crazy ideas, which she always described to Mama as "great opportunities!" Like the summer before I met Derrick, when Miss Munroe got it in her head that I should go to Free Sleep-away Camp.

"This will be a wonderful opportunity," Miss Munroe had gushed to Mama, standing in the living room with a camp brochure in her hand. "There will be all kinds of fun activities, like swimming and hayrides."

"I can't afford all that," Mama told her.

"Not to worry," said Miss Munroe with a smile. "The camp is subsidized. It won't cost you a thing."

All Mama said was "hmmph," which I guess Miss Munroe took as a yes. Because before she left, she cheerily handed Mama a typed-up list of everything I would need for camp: a sleeping bag, pillow, bathing suit, beach towel, tennis shoes, five pairs of shorts, eight T-shirts, flip-flops, and a flashlight. Then she gave Mama a Kmart voucher to pay for it all.

Mama took the voucher from Miss Munroe and stuck it in her wallet. She put the list on top of her TV, under an ashtray, where it stayed, untouched, until the Saturday morning two weeks later when Miss Munroe came back to pick me up for camp.

"Patricia, honey," she said in her singsong voice as she walked into the living room, "are you all packed up and ready to go?"

"No, ma'am," I answered.

"What do you mean?" Miss Munroe asked, sounding confused. I saw her eyes dart around the room. Her gaze landed on the camp list sitting on top of the TV, covered in spilled ashes and beer stains. She cleared her throat and turned to Mama: "Miss Williams, where are Patricia's things?"

"I wasn't able to pick up none of that shit," Mama said, staring at the ground.

"But you knew your daughter was going to camp today?"

"Yeah."

"And you couldn't get *any* things on the list?"

"Nah."

"Not one thing?"

"Nope."

Miss Munroe pressed her lips together so tight her mouth disappeared from her face. "Was there"—she cleared her throat—"a *problem*?"

"I been busy."

Miss Munroe glared at Mama for what felt like a full minute. She looked so mad, standing there with her mouth clamped shut, I thought for sure she was gonna pick up an empty Schlitz can from off the floor and throw it at Mama's head. But instead she straightened her back, grabbed me by my hand, and marched me out of the apartment.

She took me to Kmart herself, speed-walking through the store, snatching up T-shirts, towels, and flip-flops and pitching them into her cart. Mama won the battle over who would buy my supplies for Free Summer Camp, but Miss Munroe won the war. After Kmart she put me in her Buick and drove me there herself. When I came home two weeks later, Mama didn't say a word.

IN THE LIVING ROOM, I stood with my back against the wall and watched Mama and Miss Munroe like I was watching a tennis match, only I was the ball. Miss Munroe tapped her clipboard. "There is also the matter of the child's father," she said, looking up. "Patricia tells me her boyfriend is twenty years old."

Mama shot me a look.

"Have you met this young man?" continued Miss Munroe.

"Yeah, I met him."

"So you're aware of his age?"

"I don't know nothing about that," said Mama. "I don't know how old that boy is. "

"Miss Williams, as I'm sure you're aware, it's a crime for an adult to have sexual relations with a minor. My advice to you is to get the authorities involved. You need to file a complaint with the police."

"The *police*? Oh, hell, nah! I ain't talkin' to them."

"Miss Williams, the father of this baby is an adult male who has committed statutory rape."

"Rabbit," Mama said, turning to me, "did this boy rape you?"

"No, ma'am," I said. "He my *boyfriend*."

Mama turned back to Miss Munroe: "She say he didn't rape her."

"But she's a *child*!" said Miss Munroe, her voice rising. I'd never heard our caseworker yell before. Her anger startled me, but Mama just narrowed her eyes.

"I already told you," she said for the very last time. "The answer is no."

Before she left, Miss Munroe pulled me aside. She handed me a pamphlet for the Free Prenatal Clinic at Grady Hospital and a bus pass to get there. "It's important for you to see a doctor and to make sure you and the baby are healthy," she said.

"Yes, ma'am," I answered. "I'm gonna be a real good mama."

"Of course you are," she whispered, leaning over to give me a hug. I was surprised to hear her voice catch in her throat as she told me good-bye.

CHAPTER 10

Wife on the Side

I couldn't wait to tell Derrick about how I saved his ass from jail. I imagined him leaning over, putting his arm around me, and telling me I was the most ride-or-die girlfriend ever. Maybe he'd even bust out a song, like they did in the movies, singing to me the chorus of Whitney Houston's "You Give Good Love" as a sign of his appreciation. I was so excited about him pouring his thank-yous all over me that I ran to his car the minute I saw him pull up outside Mama's place the next morning when he came by to drive me to school.

I told him the whole story, barely pausing to catch my breath: "Miss Munroe wanted Mama to call the police. But I said you were my *boyfriend*. I told her it wasn't no rape!"

When I was done, I leaned back in my seat waiting for my thank-you, but instead Derrick just sat there, rocking back and forth with his thumb in his mouth. "Oh man," he groaned. "Oh man . . ."

"What?" I asked, wondering if maybe he didn't understand what I just said.

"I can't get locked up behind no shit like this, Rabbit. I can't go to jail."

"That's what I just said! You not going to jail because *I* told Miss Munroe we go together!"

"Urgh . . ." Derrick leaned his head back and made a sound like he'd just been punched in the stomach. Then he turned to me. "Get out the car," he said, suddenly.

"What . . . *why?*"

"Just go in the house," he said staring straight ahead. "I gotta bounce."

I was so stunned, I didn't move.

"For real, Rabbit. Get out the car." Derrick leaned over me to open the passenger-side door. I didn't understand why he was being so mean. I knew he had his moods—sometimes he'd snap at me for no reason, and once he grabbed me by the top of my arm for having a smart mouth—but he'd never talked to me like this before, kicking me out of his car.

"But you coming back, right?" I asked.

"Yeah, whatever. I'll see you tomorrow."

I waited for him the next day after school, but he never showed up. Not the day after, either. I called him at Fish Supreme, but he wouldn't come to the phone. I went by his sister's place, but she said she hadn't seen him. I sat on Mama's front steps watching the road for hours hoping to see his car, but he never drove by. Weeks passed and still no Derrick.

At school, I'd sit in the back of the classroom and write over and over in my exercise book, *I love Derrick, I love Derrick, I love Derrick . . .*

ONE SATURDAY MORNING almost a month after Derrick ran off, I was glued to the TV set trying to take my mind off my troubles with *The Smurfs*. I was interrupted by someone knocking at the front door. My heart jumped into my throat.

DERRICK!

I raced to the front of the house and flung open the door. But instead of my boyfriend, standing on Mama's porch was a lady I'd never seen before, dressed in jeans and a lavender T-shirt stretched tight across her belly.

"Is Rabbit home?" she asked.

"I'm Rabbit."

A look of surprise flashed across her face. "*You're* Rabbit?"

"Yeah," I said. "Who the fuck are you?"

She put her hand on her chest. "I'm Evaleen," she announced. "I'm Derrick's wife."

I stared at her, confused. What did she mean "wife"? Derrick didn't have a wife.

"Me and Derrick been married more than a year," she went on. "We live out in Decatur."

Decatur? I thought. *Derrick told me he stayed by his mama's house, around the way.*

". . . with our baby," Evaleen continued.

Baby? What baby?

"Our daughter's going on seven months now." Evaleen paused, as if to let the news sink in. "We got another one on the way." She patted her stomach. "I'm pregnant. This one's a boy. We're naming him Derrick Junior, for his daddy."

What the hell? 'Derrick Junior' was my *idea for a baby name.*

Evaleen looked at me like she was waiting for me to say something. But all I had were questions.

"How are you his wife?" I asked. "You have to be a girlfriend before you're his wife and *I'm* his girlfriend."

Evaleen sighed. "How old are you, anyway?" she asked.

"Thirteen."

She put her hand to her mouth and gasped: "That dirty dog."

For a minute the two of us stood there, staring at each other in shock. I tried to make sense of what Evaleen was telling me. I played back all the months I'd been with Derrick and thought about the times we'd had sex in his car, at the park, or in Catfish's old apartment on the dirty floor. I wondered if Evaleen was the reason Derrick never took me to his house or to meet his mama. Maybe him having a wife is why his sister always gave me the stink eye every time he brought me over to her place.

My mind flashed back to this one time when Derrick and I were sitting in his car and a girl ran past, banging on the front window. "That nigga married!" she yelled. "He married!"

Evaleen must have been putting the pieces together, too, because standing on my doorway she closed her eyes and began to pray. "Lord," she said, "please give me the strength not to kick his sorry ass for this unholy alliance and transgression."

When she was done, she opened her eyes and looked at me. "Girl," she said, "I know you pregnant. We need to talk."

I couldn't imagine what she wanted to talk about. But just as I was opening my mouth to ask her, from up the block came the familiar sound of the ice cream truck. It made its way toward us, and parked right at the curb in front of Mama's stoop. Evaleen glanced at the truck, her hand on her belly.

"You want one?" she asked, reading my mind. I was pregnant. Hell, yeah, I wanted ice cream!

I nodded yes and followed her to the truck. Evaleen bought herself a Creamsicle and handed me a Bomb Pop. Then she got right back to business.

"How far along are you?" she asked.

"Almost four months."

"Good, there's still time."

"Time for what?"

"Time for you to get an abortion. You're not even showing yet."

"Why would I get an abortion?"

"Because if you have this baby it's really gonna mess up my marriage," she explained. "I have a family. He's *my* husband so *you* need to get an abortion."

I couldn't believe I'd only met this lady three minutes ago and now she was asking me to kill my baby. In fact, I couldn't believe any of this was happening. The day before I'd been a regular pregnant seventh grader. Now, suddenly, I had the kind of problems I'd only ever seen on *The Young and the Restless*.

I started to tell Evaleen that she wasn't the boss of me—I was having my baby no matter what *she* wanted—when, from out of nowhere, Derrick drove up in his Chevy, like a bat out of hell, screeching to a stop in front of us.

"Bitch!" he hollered out his car window. At first I thought he was talking to me. But it was his wife he wanted. "Evaleen!" he yelled. "What the hell you doing? Get your ass in the damn car!" I stood on the curb with my Bomb Pop melting down my hand and watched Evaleen slide into the passenger seat. As soon as she closed the car door, Derrick punched her in the face.

I WALKED BACK INTO THE HOUSE. I wasn't ready for all this. These were grown-folk problems and way too much for me to handle. Especially since, with Derrick gone, I had no one to talk to. He had been my only friend. I wanted things back to how they used to be, when all I had to worry about was what to wear when Derrick took me roller-skating.

I lay down on the sofa and curled into a ball. With Smurfette giggling on the TV in the background, I tried to face the facts: my baby daddy was a low-down lying cheat. Still, all I wanted was for him to come back.

CHAPTER 11

It's Time

I was eight months pregnant and ready to pop when Mama told us to pack up all our shit because we were moving again. I don't know how she could predict the exact number of months and weeks she could go without paying rent before she got evicted, but she always knew when it was time to leave.

Other folks weren't so organized. They'd fall behind on their rent and then a red eviction notice would get nailed to their front door and the marshals would show up and haul all their furniture and clothes and dishes and personal items out onto the curb. Getting put out was bad for other people, but good for us because that's how we got new furniture. Mama would send us kids out to pick through the stuff the marshals had left by the side of the road, before the tenants came home to find out they'd just become homeless.

The new place Mama moved us to was in a run-down neighborhood called Vine City, and it was the worst place we'd ever lived. The apartment was called an "efficiency," which I guess is short for "one step before hell." It was a single, cramped, roach-infested room with a kitchen along the back wall and a small bathroom with a stand-up shower. It was only supposed to be for one person, but we all stayed there: me, Mama, Sweetie, Sweetie's three-month-old daughter LaDontay—named for her baby daddy, Dontay, who happened to be Derrick's younger brother—and my brothers Dre, Andre, and Jeffro when they weren't locked up. It was such a shit-hole that when Dre got out of juvie and saw how we were living, he flagged down the popo and asked them to take him back to jail.

For home furnishings we had two beat-up sofas we picked up on the street. Mama slept on the brown corduroy one. Mine was dark yellow with a pattern that looked like flowers from far away, but on closer inspection seemed to be cooking grease and body fluids. Sweetie and her baby slept on a piece of foam we laid on the floor between the sofas. During the day we stored her foam in the shower stall, which meant we had to wash up in the sink. But first we had to heat the water on the hot plate because Mama hadn't caught up on paying the gas bill, so there was no hot water.

We'd been living in Vine City for a month when I woke up in the middle of the night feeling like a thick rope was tightening around my belly. The pain came and went like waves coming in. I lay there for a while, trying to wish it away. But when it got so bad I was sure I was dying, I called out to Mama.

"Ohhhhhh . . ." I moaned. "My belly hurts."

"Okay," she said, flicking on her lamp. "Maybe you just got some gas. What it feel like?"

"I dunno."

"Like you need to doo-doo?"

"I don't think so."

"Do it feel like a regular upset stomach or more like your cycle coming on?"

"I don't know what it feel like," I said, turning my face to the wall. "It just hurts."

Mama didn't trust a doctor, so whenever something was

wrong with one of her kids, she liked to do the diagnosing herself by asking a million questions and then taking a wild guess. Over the years she'd told me I had infantigo, trench mouth, chicken pox, sour stomach, a case of the nerves, and fleas. No matter what the ailment, the remedy was always "rub some Vicks on it."

"Do it hurt in the front or the back?" she asked.

"All around."

"Like you getting squeezed real tight?"

"Yeah, like that."

"Oh girl, sound like you in labor."

Mama lifted herself half off her sofa and swatted at Sweetie, who was still sound asleep on her foam. "Get up, girl. Go to the pay phone and call 911 before your sister has her baby on my damn floor."

Sweetie groaned and turned over, but didn't make a move to get up. So Mama kicked her, hard. "Get the fuck up and go make the call!"

I rolled over onto my side. The pain had only started, but it was already way worse than I'd imagined, and I was scared about what was coming next.

"Contractions will increase in intensity and duration, signaling the impending arrival of your baby," a nurse at the Free Prenatal Clinic had tried to warn me, reading from a pamphlet called "The Exciting Days Ahead." I just stared at her because I didn't know what the hell she was talking about.

"Girl," the nurse had said, leaning forward. "This means it's gonna hurt like a son of a gun. When the pain starts coming, you find your way to the hospital."

"Ohhhhhh . . ." I moaned.

"All right!" said Sweetie, jumping up and handing her sleeping baby to mama. "I'm going! Hold your horses, Rabbit. All you gotta do is squeeze your legs together so that baby don't pop out your coochie while I'm gone."

"You tryna be a smart-ass, now?" Mama said. "You need to get a move on and go make that call. No telling how long the ambulance gonna take to get here."

It was a known fact that 911 took their sweet time showing up to Vine City. Just about everybody had a story about an uncle or brother or cousin who almost bled to death on the gotdamn sidewalk because the ambulance took their muthafuckin' time.

Lucky for me, Sweetie had a special talent with 911. She discovered it by accident one day when Dre was twelve years old and had an asthma attack after he and Andre tried to get high huffing gasoline out of the gas tank of Mama's Pink Panther.

Sweetie had run to the pay phone and screamed into the handset, "My brother can't breathe. He dying!"

Minutes later, an ambulance showed up, sirens blaring. Sweetie and I piled into the back with Dre. When we got to Grady Hospital, they wheeled our brother into the ER and Sweetie and I followed him in. While we were waiting for a doctor, a little old lady hospital volunteer in a pink smock came by handing out free sandwiches wrapped in cellophane and little containers of apple juice. She didn't even care that Sweetie and I took four sandwiches each. "We gonna hold these for our brother," Sweetie had said, and the volunteer just smiled.

From then on, whenever food got real scarce Sweetie would call 911, Dre would have a fake asthma attack, and we would all get to eat. It wasn't a regular thing. That would be abusing the system. But there are certain problems in life—like being really really *really* hungry, or going into labor—that can only be solved by calling 911. That's when Sweetie's God-given talent for sounding hysterical came in handy.

Sure enough, only minutes after my sister came back from the pay phone, two EMTs showed up at the door. That was quick, even for her.

"Someone here call for an ambulance?" asked the ambulance man, looking around the room. He was short and round. His partner, walking in behind him, was tall and skinny. The two of them reminded me of Abbott and Costello, who I used to watch with

Granddaddy on his little black-and-white TV, both of us laughing our heads off.

"That's her right there," Mama said, pointing at me. "You need to take her to Grady."

I sat up on the sofa but the two EMTs just looked at me, confused. "Dispatch said they had a call about a *birth in progress*," said the chubby one. "Right here at this address."

"Yeah," said Sweetie, pointing at me. "She's got a birth in progress. You can't see she good and pregnant?"

"Miss," said the skinny one, turning to me, "when exactly did the pain start?"

"Dunno," I shrugged. "Maybe half an hour?"

"So you're not actually giving birth at the present moment?"

"No, sir."

"Then why did we get a call about a birth *in progress*?"

"Because I told the 911 lady that the baby was coming out," answered Sweetie, who was standing on her mattress with her hands on her hips. "I said I seen the top of the baby's head. That's what I told her."

The skinny ambulance man looked at my sister like she'd just announced she stabbed an old lady in the eyeball: "Why would you *do* that?"

"So y'all muthafuckas would get here!" Sweetie practically yelled. Then she turned to me. "You WELCOME!"

The ambulance men didn't want to take me. They said I had "plenty of time," and the only reason they'd come so quick was because they were already in the neighborhood, on another call.

"We diverted from another patient because dispatch said 'birth in progress,'" the chubby one said, holding up his fingers in air quotes. "As in 'an emergency.'"

The skinny one said to Mama, "Ma'am, why don't you call again when she's further along and they'll send another unit?" Without waiting for a response, the EMTs turned to leave.

"Oh hell, nah!" Mama yelled. "You gonna take her right now."

The chubby one shot his partner a look and shook his head, like he couldn't believe somebody was asking him to do his actual job. "All right," he said, with a sigh. "I'll get the stretcher."

"I don't need all that," I said, standing up. "I can walk."

The skinny guy pushed me back down, hard. "No ma'am," he said. "It's regulation."

Soon after the chubby one came back with the stretcher, and the two of them started strapping me in. That's when it really hit me: I was having a baby.

Suddenly my heart was pounding like it was coming out of my chest. This whole thing was a BIG MISTAKE, I wanted to yell. There's no way I was gonna be able to fit this baby out my cooch. NO GOTDAMN WAY!

I looked over at Mama, who was leaning back, puffing on a Winston. I noticed she wasn't wearing shoes.

"Aren't you coming with me?" I asked, alarmed.

"Nah, girl," she said through a cloud of smoke. "You heard the man. It's not like you gonna have the baby tonight."

"But Mama . . ." I said, panicking. "I need you. Please, can you come? *Please?*"

"What you need me for?"

"'Cause I'm scared."

"C'mon now," she said. "Girl, there ain't nothin' to be scared of. Your sister had a baby and she didn't have no trouble at all. Ain't that right, Sweetie?"

I looked over at my sister, but she just shrugged. She couldn't tell me it was going to be fine because for weeks she'd been describing the miracle of her childbirth. It sounded like the worst X-rated horror movie ever made: *"Rabbit, for real, it feels like your insides is being ripped out . . . like you got an alien in there and it's tryna to kill you . . . when the doctor cuts you down there to let the baby head out you won't even feel it cuz you already gonna be in so much pain . . . and you know you gonna shit yourself, right? Yep, you gonna doo-doo right on top of your little baby's head. . . ."*

"You ain't got nothin' to worry about," Mama said again. "You don't need me tonight."

I turned away from her to face the wall, trying not to cry. I was so mad at myself for begging her to come with me. I should have known Mama would never say yes. If there is one lesson she'd taught me in life it was not to ask her for shit: not food, or clothes, or to comb my hair. But I had been fooled, because lately things had been different.

Ever since school let out for the summer, Mama and I had been together from dawn till dusk, sitting in that hot-ass efficiency watching all her shows: *The Price Is Right, The Young and the Restless*, and *As the World Turns*. One day she looked at me and said, "Girl, you big as a house! That's just how I was when I was pregnant with you." She even put her hand on my belly and felt the baby kick. It was the closest we'd ever been. If felt like we were bonded. Like she was gonna be there for me, for real.

Instead all she said was "No use in both of us going to the hospital," as the ambulance men lifted up the stretcher and carried me out the door.

OUTSIDE, I FELT THE NIGHT AIR against my face. I closed my eyes and promised myself that I would never ever ever EVER ask Mama for shit again as long as I lived. I gripped my belly and squeezed my eyes shut to keep the tears from falling. When I opened them again, I was inside the back of the ambulance.

I looked around and was startled to see, on the bench against the wall, a balding, middle-aged man wearing dirty sneakers with no laces. He had a bandaged right shoulder and his faded T-shirt was ripped and covered in blood.

"Who the fuck are you?" I said. I tried to sit up, but I was strapped to the stretcher. "What the hell?"

"Miss," said the chubby ambulance man. "I'm gonna need you to calm down."

"But who is *he*?" I pointed at the man covered in blood.

"Well, miss, as I previously informed you, we were tending to this gentleman and his injuries when we got the call about a woman giving birth. Now if you just lay back down, we're gonna take the two of you directly to the hospital."

I turned to the bleeding man. "What happened to you?"

"Nigga cut me," he answered, with a shrug of his good shoulder. I could smell the liquor coming off him like he'd been swimming in a vat of Granddaddy's moonshine. He looked from my face to my belly. "You pregnant?" he asked. "Or is you just fat?"

"What it look like?" I said, suddenly irritated. "CAN'T YOU SEE I'M HAVING A GOTDAMN BABY?"

"Okay, okay, okay . . ." He put up his hand like he was stopping traffic. Then he leaned back and stared at me like he was trying to get his eyes to focus.

"Now, miss," he said, pointing his finger in the air. "Let me tell you somethin'. Babies are a gotdamn blessing, ya heard me? I got four children of my own: Rodney, Samuel, Joseph Jr., and Joleen, my little baby girl. She 'bout three, maybe four months old. Can't be no more than six months. Maybe a year, give or take. Hell, it don't matter! That baby is the prettiest little baby you ever seen. I enjoy the shit outta them kids. So I just want to say congratulations to you, miss!" He swept up his arm like he was holding a Bumpy Face bottle and making a toast. "Congratulations!"

"I mean that," he added. "It's a gotdamn beautiful thing."

Just then another contraction hit. I grabbed my belly and moaned, *"Ohhhhhhh . . ."*

"Now that shit gotta hurt," said the drunk after it passed. "You ain't got nobody ridin' with you to the hospital?"

"No."

"You want to hold my hand?"

"Nope."

"All right then. Good luck."

"Yeah, you too."

CHAPTER 12

Baby Formula

My daughter was born at 1:07 P.M. on August 9, 1986, weighing eight pounds, fifteen ounces, with big brown eyes and a full head of hair. "You so pretty," I whispered when the nurse placed her in my arms. I named her Ashley, after the beautiful Ashley Abbott from *The Young and the Restless,* who was having an affair with an older, married man, just like me. Ashley Abbott and I had almost the exact same life problems. The only difference was her married man was Victor Newman, the billionaire owner of the worldwide cosmetic and real estate conglomerate Newman Enterprises, while mine worked the fryer at Fish Supreme.

Derrick didn't have Victor Newman money. But that didn't stop me from having big dreams about the life of luxury I wanted for our baby. Most of my fantasies involved white-lacquer furniture. The kind I'd see when I went with Mama to Carson's, a buy-now-pay-later furniture store where Mama was paying off her nineteen-inch color TV.

Every month, rain or shine, as soon as she cashed her welfare check, Mama would get in the Pink Panther and drive over to Carson's with her ten dollars in hand. Mama might fall behind on the rent, the electric, or the gas, but there was no way she was gonna miss a payment and let Carson's repo her TV. While she argued with the cashier about how much she still owed, I'd wander over to the Baby Room to take a look at floor displays of the most beautiful furniture I'd ever seen.

I had my eye on a shiny white crib that came with a matching dresser. I'd run my hand along the rail and imagine my baby luxuriating in that crib, dressed in a pink Adidas tracksuit with a pink baby bow around her head and tiny British Knights tennis shoes that I'd clean with a toothbrush. To me, Carson's Baby Room represented everything good in life: a clean home, nice things, a mama who cares. It reminded me of sitting at Granddaddy's bar, my eyes glued to his black-and-white set, watching *Leave It to Beaver,* my favorite show. I noticed every little detail in that TV house, from the checkerboard curtains in the kitchen window to the way everybody was always smiling and nobody ever got mad.

I didn't know how I was going to get a *Leave It to Beaver* life for my baby, but I knew it's what I wanted. I could see it clear as day: Me, Derrick, and our little girl, smiling in our home full of gleaming white furniture. I imagined it so much it felt like it was real.

MR. JOHN BROUGHT MAMA AND SWEETIE to visit me the afternoon Ashley was born. The next day Derrick showed up. I'd only seen him a handful of times since the morning he kicked me out of his car for saving his ass from jail. But this was the day I'd been dreaming of, when he'd come back and the three of us would be a family.

My heart pounded with excitement as I watched Derrick take our baby in his arms, grinning like a fool. He insisted Ashley looked just like him. "She got her daddy's eyes!" he exclaimed. He stared at her for a good twenty minutes. Then he got bored and handed her back to me. "I gotta bounce," he said.

"But you'll come back tomorrow, right?"

"Yeah, I'll be back."

I took a breath and started pulling myself out of the hospital bed, careful of the stitches in my cooch.

"Where you going?" Derrick asked.

"To walk you to the elevator."

"Nah, you don't need to do all that," he protested.

But I was already shuffling out the door.

I headed down the corridor dragging my IV pole behind me. My titties were leaking milk all over the front of my hospital gown and my hair was standing up like I'd just put my finger in an electric socket.

Derrick was two steps behind me when I turned the corner to the elevators, so I saw the girl standing in the hallway before he did. She was leaning up against the wall dressed in a matching denim skirt-and-jacket set decorated with pink and purple rhinestones. When Derrick came around the corner behind me, she gave him a little wave, smiling at him like I wasn't even there.

Maybe it was the baby hormones messing with me, but seeing that girl looking so put together when I felt so busted was like a donkey kick to the stomach. I turned to face Derrick. "Who the hell is that?"

"Don't start trippin'."

"Who is she?"

He paused. "That's Celeste."

I looked from Derrick to Celeste. She looked at me and back at him.

"We just friends," Derrick said. "That's it." By the expression on her face, this was news to Celeste. I didn't believe him, either.

Suddenly I could feel hot tears flooding my eyeballs, but there was no way I was gonna let Derrick see me cry. "Fuck you," I said. I grabbed my IV pole and headed back to my room. "You nothing but a dirty dog."

I bawled like a baby that night. I couldn't believe I'd been so *stupid,* caught up in a fantasy of all the nice things I wanted for my baby when really I didn't have shit. In a few days I'd be taking Ashley home to Mama's nasty efficiency with the dirty cast-off sofas, no hot water, and the foam mattress on the floor. I didn't have diapers or formula, blankets or clothes. I didn't have little girl dresses or pink bows for her hair. I didn't have a single baby bottle or wipes or a clean washcloth. I didn't even have Derrick.

That's when it really hit me. I was gonna have to figure this shit out all by myself.

EARLY THE NEXT MORNING, before the sun came out, I got to work. I snuck out of my hospital bed and made my way down the hallway to the metal supply rack I'd noticed sitting beside the nurses' station. When I was sure nobody was looking, I grabbed a stack of diapers and five baby T-shirts and shoved them under my gown. Back in my room, I hid them in my covers. Then I went back for more.

It turns out nurses are almost as easy to steal from as drunks. They get distracted by every little thing—some juicy gossip, a doctor bossing them around, a patient hemorrhaging down the hall— that's when I'd make my move. I swiped stuff out of an unlocked supply closet, from the counter at the nurses' station, and from the carts the nurses wheeled into my room When they discharged me from the hospital, I left with two trash bags stuffed with dozens of bottles of pre-made Enfamil, baby clothes, and diapers. I thought I'd be okay for a while. But Ashley was only three weeks old when the diapers ran out.

"What am I supposed to put the baby in?" I asked Mama, holding up my half-naked daughter. "The Pampers is all finished."

"Girl, use your damn head," Mama said. She took Ashley from my arms and showed me how to use an old T-shirt for a diaper, pinning it closed with a safety pin. "See? Granma's baby is good as new."

Mama got thirty-four dollars a month extra on her welfare for having a new baby in the house. She gave it to me to buy baby formula. But the money was never enough. When Ashley was hungry and wouldn't stop crying, I went to the Vine City corner store and slipped two containers of Enfamil down the front of my shirt. When the Enfamil was finished, I gave Ashley watered-down Carnation Evaporated Milk. When I ran out of that, Mama said, "The baby's old enough for table food." So I took a bite of my ketchup sandwich, chewed it up, spit it into my hand and fed it to my baby like she was a little bird. She was three months old.

One day Ashley was crying her head off. Mama told me to go to the pay phone on the corner and call Derrick. "That piece of shit you had a baby by is supposed to be helping you," she said. "Tell him to bring you some Pampers and milk."

I dialed the number for Fish Supreme and could hear Derrick's coworker telling him he had a call: "Brotha, it's your baby mama on the phone. . . . C'mon, man, how the hell am *I* supposed to know 'which one'?"

Derrick promised he'd come by with some money, but he never showed up. Instead, when I opened the front door later that afternoon, there was Celeste, holding two Kmart shopping bags filled with diapers and baby clothes.

"Derrick told me to bring these," she said, handing me the bags. I couldn't believe he'd sent her. I was even more pissed off that she had the nerve to show up at my door dressed in a cute purple skirt and vest set when I had on a stretched-out T-shirt decorated with baby puke.

"I don't want this shit," I said, grabbing the bags out of her hands and throwing them past her, into the air. Little baby dresses and diapers flew out of the shopping bags, landing all over the dusty yard. Celeste just shook her head: "Girl, you better pick that stuff up. You know you need it."

"You don't know what *I* need!" I yelled. "What *you* need is to get the fuck outta here!" I slammed the door in her face and waited inside, my back pressed to the door, until I thought she was gone. Then I went to the yard and started gathering up the diapers, booties, socks, and little pink dresses. My eye caught on a pink bow fixed to a elastic headband. It looked just like the kind of baby bow I'd imagined Ashley wearing all those times I daydreamed at Carson's. I bent down to pick it up. When I looked up I noticed Celeste sitting behind the wheel of her Camaro, watching me on my hands and knees picking baby clothes out of the dirt.

ON A COOL AFTERNOON, when Ashley was almost six months old, Miss Munroe came by to check on me. We sat on Mama's stoop and she took my hand in hers. "I'm concerned," she said, her face serious. "How are you managing?"

"I'm good," I replied.

She squeezed my hand. "Really, Patricia? Are you?"

Suddenly I was sobbing, with tears and snot rolling down my face. I told her Sweetie's daughter LaDontay cried all night, and Ashley and I couldn't get any sleep. I told her that Mama wasn't helping me and Derrick wasn't around. "I'm trying," I cried. "But I keep running out of everything."

Miss Munroe patted me on my back, and handed me a tissue from her purse. "I think there may be something we can do," she said. Miss Munroe explained that I could get my own public assistance—two hundred and thirty-five dollars in welfare, plus food stamps—if I became an emancipated minor. "It means your mother would no longer be legally responsible for you," she added. "The benefits would go directly to you."

She helped me fill out the paperwork, and a few months later I packed up all my stuff in trash bags and dragged Mama's dirty yellow sofa to my very own place, an efficiency across the yard from Mama's. It wasn't the luxury accommodations I dreamed of, but for the first time in years when I turned on the faucet the water was hot.

DERRICK MUST HAVE HEARD I moved out from Mama. Because I'd only been in my new place a couple of weeks when he started coming back around. I tried to be strong. "I don't need your cheating ass no more," I said, standing in the doorway with Ashley on my hip. But he cocked his head to the side and gave me a sly smile.

"C'mon, Rabbit," he said. "You ain't gonna let me see my baby? I thought we was a family."

Before I knew what was happening, he reached out his hand and touched my arm. It was gentle, like how you'd pet a sick puppy, but I felt like he'd set me on fire. My neck grew hot and my palms began to sweat. Then he was leaning forward, whispering, "You know I love you," with his warm breath in my ear. I hadn't felt good in so long. With my heart pounding, I opened the door and let him in.

The nurse at Grady Hospital had given me a box of condoms when they sent me home with Ashley. "You're young," she'd said matter-of-factly. "You can go back to school with one baby. You don't need to have another."

I didn't want a second baby. But every time I handed Derrick a condom, he laughed in my face. "Look what I'm working with," he said, waving his hand in front of his wiener like he was a *Price Is Right* spokesmodel and his junk was a brand-new washer-dryer set. "No way that thing's gonna fit."

Six months after Ashley was born, I was pregnant again.

I tried to handle the situation as best I could. I knew I'd need more money, so I got some fake ID that said I was eighteen—old enough to work—and got a job waitressing the overnight shift at the Huddle House. I paid Mama ten dollars to watch Ashley. But going to school during the day and working all night, and being pregnant, was just too much. I dropped out of eighth grade. The next month, I got fired from the Huddle House for stealing five dollars out of the till. All the waitresses were doing it, but I got caught.

My son Nikia was born in November. I was fifteen with two

babies under the age of two. I wanted to look for another job but Mama said Ashley was too damn big for her to watch anymore. My welfare wouldn't stretch the whole month. I started falling behind on the rent and the bills. It felt like I was drowning.

One night I lay on my yellow sofa with both my babies knocked out on top of me. Ashley started fussing first, that set off Nikia. Then both of them were crying their eyeballs out. I didn't know what else to do so I closed my eyes and called on God: "Dear Heavenly Father, I know I haven't been to church in a while, but I really need your help . . ." I prayed for strength and guidance. But mostly I prayed for money. "Please God, just enough to pay the rent, buy some Pampers, get my hair done . . ." I knew if God didn't come through soon, I was gonna be out on the road collecting aluminum cans for change.

I was hoping God would deliver me a paper bag full of cash. That's how I pictured my blessing coming down. But instead He sent Derrick knocking on my door.

Derrick walked into my place with a big-ass smile. When I asked him why he was so happy, he reached into his pocket and pulled out a wad of cash. It was more money than I'd seen anybody have since the roll of bills Granddaddy kept down his pants. Derrick peeled off a bunch of tens and twenties and handed them to me. "Here's a little change for you and the babies," he said.

I looked at the bills in my hand and back at Derrick. "What happened? Fish Supreme gave you a raise?"

"Nah," he said. "I quit that bullshit."

"So where'd you get this money from?"

"I'm working for Markee now," he said. "Selling that shit."

It was the spring of 1988 and Derrick's cousin Markee had hired Derrick for his booming business distributing the fastest-selling product to ever hit the hood: Derrick was selling crack.

CHAPTER 13

Hustlers and the Weak

Right from the beginning it was clear to me that crack divided the world into two groups: sellers and smokers, the hustlers and the weak. Before Derrick started working for Markee, I'd only heard about crack in N.W.A lyrics. It was a West Coast thing, like gangbanging or wearing slippers with tube socks. But the minute it touched down in Atlanta, crack spread like wildfire. I started seeing signs of it everywhere, from zombie-looking addicts trolling the streets to broke-ass niggas like Derrick suddenly getting paid.

The first thing Derrick bought himself was some bling: a gold nugget ring, a thick herringbone necklace, and a gold-plated watch. Then he upgraded me and the kids, moving us out of the busted efficiency in Vine City to a nice one-bedroom across town. Derrick took care of everything; he paid my rent, electric, gas, and water bills. He got me a touch-tone, wall-mounted house phone with a spiral cord so long it stretched all over the apartment. And he bought me a queen-size bed covered in red satin sheets that he picked out himself.

Derrick had so much money that no matter how much I asked for—fifty dollars, one hundred dollars, three hundred and fifty for the rent—he would just reach into the black fanny pack he had strapped around this waist and hand it over. "I got you," he'd say. When he had too much money to fit in his fanny pack, Derrick stored his bills in the trunk of his car in a brown paper sack inside a sneaker box. Sometimes business was so good it was like he was trying to give his cash away. "Here's a little extra," he'd tell me, popping the trunk and handing me a stack of paper. "Take you and the babies shopping."

Derrick's new job beat the hell out of Fish Supreme. After struggling to take care of Nikia and Ashley, suddenly I had everything I needed. Derrick gave me money to get my hair done, buy McDonald's, and pay for the babies' clothes. Thanks to Derrick, I didn't need to worry about a thing. Then one day, right when my rent was due, he disappeared.

I looked for him everywhere. I went by his cousin Markee's place, and searched for Derrick on the corner where he hustled. I checked for him at Jellybean skating rink, and over by his sister's house. I blew up his pager with "911," but he never called me back.

After two days of not being able to find his ass, I picked Nikia up from the sofa, took Ashley by the hand, and got on the city bus to go to Derrick's apartment, where he lived with his new girlfriend, Poochie.

Derrick moved in with Poochie after his wife Evaleen finally cut him loose for having too many extramarital kids. She had put up with his cheating when the only baby-on-the-side he had was Ashley. But when Derrick got me pregnant with Nikia, and at the same time got Celeste pregnant, Evaleen kicked Derrick to the curb. Then Celeste stopped messing with him, too. I thought for sure Derrick was gonna move in with me after those two told him "bye." But instead Poochie came outta nowhere and jumped the line.

It hurt my feelings that Derrick wanted to live with Poochie instead of me. But he told me Poochie "don't-mean-a-thing" every time he came over to see me. He kissed my neck and told me I was special. I believed him because I wanted it to be true.

At his apartment, I knocked on the front door and waited for Poochie to answer. "Hey, Rabbit," she said. "How you doin'?"

When I told her I was looking for Derrick, her eyes got wide. "Nobody told you?" she asked, shaking her head. Derrick had gotten busted. The cops had been watching his ass and caught him with money in his trunk and dope in his fanny pack. They were holding him at Fulton County Jail. "He don't even have a bond yet," said Poochie. "No telling when he'll get out."

All that night and into the next day, I lay in bed staring at the ceiling, trying to get a plan together. How was I supposed pay my bills without Derrick? I had no money, two kids, no job, and a three hundred and fifty dollar a month apartment I couldn't afford on two hundred and thirty-five dollars in welfare. I had just gotten used to not having to worry every second of the day, and now I was even worse off than before. I lay in bed, tossing and turning, trying to figure something out. When the answer finally hit me, it was so perfect I couldn't believe I hadn't thought of it before. I got out of bed and called my best friend, Stephanie. She had a car and I was gonna need a ride.

"YOU SURE YOU KNOW WHAT YOU'RE DOING?" Stephanie asked. The two of us were sitting in her Camry outside my apartment, with Ashley and Nikia in the backseat sucking on Blow Pops to keep them quiet. Stephanie had promised to drive me where I needed to go, but instead she was giving me the third degree.

"You sure you got this?" she asked again.

"Yeah."

"So you don't want to ask Shine?"

"Nah," I said. "I'm good. I seen Derrick do it a million times."

"Okay . . ." Stephanie turned the key in her ignition. But I could tell by the look on her face she wasn't convinced.

I tried to sound confident: "I *got* this."

Stephanie and I had been friends ever since the day I saw her standing at the corner pay phone. She was wearing red pants, matching red flats, a yellow top, and big-ass door-knocker earrings. "I like your outfit," I said. She was so vain, that's all it took for us to hit it off.

Stephanie was twenty-two—seven years older than me—with three kids and another on the way. Most girls in her situation would have been struggling, but Stephanie had a very strict "no broke niggas" policy, which meant she and her kids were always taken care of. Her drug-dealing boyfriend, Shine, wasn't a low-level corner boy like Derrick. He was big time, slinging dope over at Techwood Homes, a huge complex of low-rise public housing apartments filled with folks looking to forget their troubles or, as Shine called them, "good customers." Shine was the one who bought Stephanie her car.

"So you gonna spend your whole welfare?" she asked as we pulled away from the curb.

"Yeah."

"All of it?"

"Yeah," I said again. "Plus the fifteen dollars you're gonna lend me."

"Okay," she said, laughing. "You better not fuck this up."

The plan was simple. I had $235 from my welfare check that I cashed at the Super Saver, plus fifteen dollars from Stephanie. That was just enough for me to buy a quarter ounce of dope from Markee. If I chopped that quarter into fifty rocks and sold each rock for ten dollars, I would make five hundred dollars, enough to cover my rent with a little extra to spare.

I'd never seen a girl selling dope before, but how hard could it be? I'd been out with Derrick when he worked his corner and, by the looks of it, you just had to stand around and wait for business to come to you.

We drove over to Markee's mama's house in Ben Hill. She lived on a quiet street where folks kept their porches tidy and nobody dumped their old refrigerator in the yard. It didn't look like the kind of place you'd find a stash house, but I guess that was the point. Stephanie pulled over to the curb and I paged Markee. Somebody in the front room of his mama's house pulled back a corner of the window curtain. I gave a nod and Markee came right out.

"Hey, Rabbit," he said, waving to Nikia and Ashley in the backseat. "The kids are getting big. That little boy look just like his daddy."

"Yeah, he sure do," I agreed. When I told Markee what I'd come for—to buy myself a quarter ounce—he raised his eyebrows.

"You got money?"

"Yeah, I got it right here."

He held out his hand like we were gonna shake, and I slid the bills into his palm. Markee glanced up and down the street, checking for cops, then went back inside. When he returned to the car, he slid me a white disc, a little bigger than a silver dollar, neatly wrapped in a Ziploc bag, folded over and stapled shut.

"Be careful," he warned. "I mean it, Rabbit. You got those kids to take care of. Don't do nothing stupid."

STEPHANIE AND I HAD ONE MORE STOP TO MAKE, at a convenience store called the BusStop. From the outside it looked like a regular corner store, with soda, chips, toilet paper, and tube socks hanging in the dusty window. But when you stepped inside, the place was like a Walmart for crack dealers. Underneath the glass counter and on shelves behind the register was everything you could ever need to cook, bag, sell, and smoke rock: triple-beam scales, baking soda, glass pipes, and sacks in every size, from one-inch dime bags to one-gallon Ziplocs. I bought a bundle of a hundred small sacks, a pack of Pace single-edge razor blades, and a box of sandwich bags. Then we headed back to my place.

At home, I gave Nikia a bottle and sat him on the sofa beside Stephanie, who was busy flipping through channels, while I got to work. The trick was to cut that piece into fifty rocks, all of them about the size of my pinkie fingernail. If I chopped them too big, I'd lose my profit. If the rocks were too small, nobody would buy them. I pulled a dinner plate from the cabinet, set it on the kitchen table, and unwrapped my product. With a razor blade in one hand, I turned the piece over, trying to figure out where to make the first cut.

"Don't make them rocks too big, girl," Stephanie said, looking over from the sofa.

"Yeah, I got this," I lied. I slid my razor blade and chopped my first rock.

It took me almost an hour to cut up that quarter. Nikia finished his bottle and fell asleep while Ashley stayed busy pushing her little plastic shopping cart around the apartment and filling it with the fake fruits and vegetables that were all over the floor. She rolled her buggy into the kitchen and tried to hand me a plastic apple from her cart, but I told her, "Mama's working," and shooed her away.

"Forty-two, forty-three, forty-four . . ." I counted when I was done, pushing each rock to the side of the plate with the razor. There were only forty-eight rocks in all; I was two short.

"I knew you was gonna do that!" Stephanie called from the other room. "That just cost you twenty dollars."

"Don't worry about it," I called back. "I wasn't about to spend it on you."

I bagged up the rocks, put them in a sandwich bag, and shoved them in my purse, and we all got back into Stephanie's car. We dropped the kids off at her mama's place. Then Stephanie and I headed to Techwood Homes to try and make some money.

We found a spot not far from a chili-dog truck that hadn't already been claimed by a dealer. Stephanie pulled up to the curb and started fiddling with her sound system. She was listening to her favorite, Salt-N-Pepa, on repeat. Everybody said Stephanie looked just like Salt, with her stacked bob pinned back on one side, long on the other. I guess she felt it was her duty to memorize all the lyrics.

I got out of the car while Stephanie danced in her seat, rapping along, *"Ooh baby baby. Bay bay bay baby, get up on this!"*

It felt like a long time that I was leaning up against the hood of her ride waiting for something to happen. Finally I saw a scrawny dude walking quickly toward me, his T-shirt tucked tightly into the waist of his high-water jeans. "You looking for something?" I asked.

"Yeah, lemme get a twenty," he said.

"I only got dimes."

"That it?"

"Yeah, man. It's real good, though."

"Aiight, lemme get two." He slid a couple of crumpled ten-dollar bills into the palm of my hand. I reached into my pocket, pulled out two dime bags and passed them back. It was that simple.

It took me three nights at Techwood to get rid of that first quarter. With the cash I made, I paid Stephanie back her fifteen, got my hair done, and took the kids to eat. I was about to take care of the rent when I had a thought: *Why spend all this money when I can double it instead?* So Stephanie and I put the kids back in the car and took another ride out to Markee's mama's place. I scored another quarter, bagged it up at my kitchen table, and headed back out to Techwood. When that dope was gone, I did it again.

Making my own money felt good. For the first time I didn't have to depend on Mama, Derrick, or welfare. Buying, chopping, bagging, and selling, it was all on *me*. I was in charge. A month after I started serving at Techwood, Markee bailed Derrick out of jail. I could have gone back to the way things used to be, with Derrick making all the money and me taking care of the kids. But it was too late for that. Once I got a taste of doing for myself, there was no way I was ever going back.

CHAPTER 14

Night Crawlers

The first dealer I heard about getting killed at Techwood was Silky. He was popped execution style with a bullet through his head. The next week another dealer went down, shot twelve times, his body dumped in the woods. Then a dope boy at Harris Homes was shot working his corner. Pretty soon all anybody talked about were the drive-bys and gunfights and corners getting shot up like the Wild West.

Maybe it wouldn't have been so bad if it was just Atlanta dealers firing at each other with handguns. But all that crack money brought gangbangers to town. The Miami Boys invaded Atlanta in the late 1980s like rats on a garbage pile. At first it looked like there were only a few of them, then suddenly they were everywhere, armed with semiautomatics and submachine guns that could tear up a whole block in seconds.

We could spot the Miami Boys a mile away. They wore Timberland boots even in the summertime, and tan-colored coveralls—like what a car mechanic would wear—that everybody called "trap suits," because in Atlanta, anyplace where you can buy drugs is a "trap." Plus, they all rocked gold teeth, top *and* bottom, sometimes decorated with diamond studs that spelled out MIAMI. Atlanta dealers might sport a single gold tooth. But a diamond-encrusted grill? We'd never seen anything like that before.

The Miami Boys were trying to take over all the projects: Capital Homes, University Homes, Harris Homes, and Techwood. I knew they didn't care about a small-time nobody like me. But still, they brought so much murder and mayhem to Techwood, I was scared from the second Stephanie and I pulled up to my spot in front of the chili-dog truck until the minute we left to go home.

WE HADN'T BEEN OUT by the chili-dog truck for more than an hour when we heard shots firing in the distance, *Pop! Pop! Pop!* And then more gunfire, this time *rat-a-tat-tat-tat* like a jackhammer.

I jumped into Stephanie's car. "Let's get out of here," I said. It was the third time that week that I'd had to leave before I'd made any money.

Stephanie drove out of Techwood and started heading toward her mama's house. Prince's "Sign o' the Times" was playing on the radio. She turned up the volume and sang along.

"In France a skinny man died of a big disease with a little name. By chance his girlfriend came across a needle and soon she did the same . . ."

We drove along Griffin Street and past the duplex where Mama cooked on a grill in the yard so everybody could see; we passed Booker T. Washington High School, where I would have been in tenth grade if my life had gone a different way; and past Catfish's apartment, where Derrick and I had done it on the dirty floor. These streets were my whole world, I realized. Anything else I knew I'd learned from TV.

Stephanie stopped singing and turned to me. "You think Prince is gay?" she asked. "Shine says for sure he's gay, but I don't know. That little dude looks hella acrobatic, though. I bet he gets *real* freaky. What you think?"

I couldn't believe Stephanie was asking me this bullshit. I guess having a money-making drug-dealing boyfriend like Shine gave her all kinds of free time to think about nothing. I didn't have that kind of life; I had *real* problems. Like how was I gonna make rent money and feed my kids without getting shot?

DRIVING THROUGH THE NEIGHBORHOOD THAT NIGHT, my mind flashed back to a time, years earlier, when I had a regular job, with regular-ass people who didn't show up for work armed like they were going to war.

Before I had my babies, when I lived with Mama on Baldwin Street, our neighbor, Miss June, had taken me with her to work at a big warehouse across town, where she had a gig as a day laborer filling cardboard boxes with Care Free Curl Activator. I made thirty dollars a day, minus the ten dollars I paid Miss June for driving me to the warehouse and lying to the foreman about my age, telling him I was sixteen—old enough to work—when really I was just thirteen and pregnant. I only worked a few weeks before Mama moved us to Vine City. But I remember the job wasn't all that bad. At least it was safe.

Stephanie wasn't far from her mama's house when I asked her to turn the car around. "I want to stop by Miss June's real quick," I said. She pulled a U-turn and stopped at the big white house with the wide front porch that Miss June and her husband had lived in for as long as I could remember. Miss June was like the kind of mother I would see on TV. She cooked and cleaned and went to church every Sunday. She always had a kind word and a cool drink for me whenever I came by. I liked Miss June, but I hadn't been by to see her in years.

"My goodness!" she exclaimed, when she opened her front door and saw me standing there. "Girl, it's been too long." Miss

June waved me to the kitchen and I followed her down the long hallway to the back of the house. She had an apron tied around her waist, and walked with a limp because part of her left leg was missing. Her husband, who was ex-military, had shot it off years before. Miss June said it was an accident. But that's what everybody said when they got shot by the same person they share a bed with.

"How's that sweet little baby of yours doing?" she asked, pouring me a glass of sweet tea.

"Actually, I got two babies now," I said. "A girl and a boy."

"Lord have mercy! Rabbit, *two*?"

"Yes, ma'am."

"Both of them by that same boy you was going with?"

"Yes, ma'am."

She shook her head. "Never much cared for him. Not one bit. But I'm gonna send up a prayer for you and your precious babies."

Miss June had seven children, six of them boys. I used to come by her place all time to hangout with Petey, her youngest. Sitting in her kitchen, it struck me that the last time I'd seen Miss June was right after Petey's funeral. He'd been shot dead, at fifteen, by a police officer who said Petey fit the description of somebody they'd been looking for. All those hours Miss June had spent on her feet packing boxes of Care Free Curl Activator, what good had it done her? She'd still come home to find the police on her front steps telling her they'd made a mistake and killed her baby boy. Suddenly, it felt like nothing was safe, not hustling at Techwood, not having a regular job. If Miss June, in all her goodness, wasn't able to keep Petey safe, what chance did I have? The thought of it made me want to lay my head down on Miss June's kitchen table and cry.

"Rabbit?" Miss June said, interrupting my misery. She was standing by the window, looking into her back yard. "You mind doing me a favor? Go out back and tell Duck to come inside and get something to eat. I don't know what he's out there doing."

DUCK WAS MISS JUNE'S OLDEST BOY. His real name was Tony, but I never heard anybody call him that. When I pushed open the back screen door and stepped onto the porch, I could see him standing on the far side of the yard, by the fence. He had his hands in his pockets and he was staring onto Baldwin Street, which ran behind Miss June's house.

"Hey, Duck," I said, walking over to see what he was looking at.

"Hey, Rabbit," he answered, quickly glancing my way, before turning back to the street. I couldn't imagine what had grabbed his attention. Nothing ever happened on Baldwin except folks sitting on their porches to catch the night breeze. All I could think of was maybe a dog had gotten hit by a car, or a couple was having a fight in the middle of the road. But as I stepped up beside Duck and peered through the fence, I froze.

The block looked like a scene from a zombie movie: there were halfway-dead-looking junkies roaming up and down the street, itching and scratching, stooped over and scanning the ground. I knew they were searching for a stray piece of rock they hoped they might find. At Techwood I'd once seen a crackhead on all fours crawling on the sidewalk, feeling around the pavement for some imaginary crumbs. That's what the comedown from a crack high did, it brought folks to their knees.

"What's going on?" I asked Duck.

"These crackheads all looking for a hit," he said.

"Where'd they come from?"

"Around here, I guess. Probably some of them walked over from Harris Homes."

Duck was ten years older than me, and he looked just like somebody's daddy. He was dressed in a bright Hawaiian shirt tucked into his carefully ironed jeans. But the crackheads must have been so blinded by their hunger, because they didn't seem to notice that this was not drug-dealer attire. As we stood by the fence, they kept coming over, asking Duck if he was holding.

"Yo, you got anything?" asked a guy in a torn Falcons jersey.

"Nah, man," said Duck.

"Got a dime?" asked a woman shuffling down the street in bedroom slippers.

"Unh-uh."

A middle age man approached us, holding the hand of a boy who couldn't have been more than three years old. "Y'all got any of that butter?" the man asked Duck.

Duck smiled at the child, then said to his father, "Man, I ain't got shit."

When they were gone, Duck turned to me: "You see this shit? All these crackheads and nobody out here selling *nothin'*. If a brother had some dope, he'd be getting paid tonight."

I reached into my pocketbook for my Ziploc bag. "Duck," I said. "I got some right here."

CHAPTER 15

Partners in Crime

Duck and I sold those fifty rocks on Baldwin in a smooth fifteen minutes. It would have taken me days to sell that much at Techwood. Spending all that time running from shootouts was obviously keeping me from living up to my full dope-dealing potential.

"Man, that was *crazy*," Duck said, shaking his head. "They was like kids at a candy store."

"That's how they do," I told him. "The minute they finish smoking that shit, they ready for more."

Duck didn't say anything for a while. He just looked out onto the street with his arms crossed in front of his chest, shaking his head "no" to the steady stream of crackheads who kept coming by looking for a hit.

Duck turned to me: "How much that package cost you?"

"Two fifty."

"And how much you reckon we just made?"

"Five hundred."

Duck let out a long whistle. "You got any more?" he asked.

"Not on me."

"But can you get some?"

"Yeah."

Duck looked back onto the street, nodding his head slowly and rocking back on his heels. I could tell he was thinking something over, but I was still surprised by what he said next. "How about you bring me another package tomorrow and I'll sell it for you?" he offered. "What would you pay for something like that?"

I couldn't believe what I was hearing. Duck, the oldest and corniest of Miss June's seven sons, was asking to be a corner boy. Right away I realized if he was serious it meant I wouldn't have to go back to Techwood.

"Twenty off a hundred," I said, offering him the standard corner-boy cut: for every hundred dollars he sold, I'd pay him twenty.

"All right," he said. "I'm in."

THE NEXT MORNING, Stephanie drove me to Markee's. I bought another quarter, took it home, chopped and bagged it up. Then I went over to Miss June's house to hand it off to Duck, just like we'd arranged.

He was leaning up against the fence in a pale green short-sleeve button-down shirt and sensible sneakers. The minute I saw him, I began to wonder if this was going to work. He looked like a school principal, not a drug dealer. The night before he'd served in the dark, and you couldn't really see him. But in broad daylight it was obvious Duck didn't look the part. I handed him the package then hung back for a while to see how things would go. That's when I discovered there was another problem, Duck was using all kinds of *manners*.

"Thank you," he said with a polite smile, every time a crack-head slid him some money. At Techwood, corner boys would blast their music, talk shit, and fuck with their customers. I once saw a dealer throw a rock into the bushes and yell to some crackheads nearby, "Go find that shit!" just for laughs. By comparison, Duck sounded like he was working the customer service counter at the Kmart. I wasn't sure junkies would go for his corny-ass look and all this hospitality. But it turns out they loved it! Duck quickly got himself customers, repeat customers, and word-of-mouth customers.

"Crackheads is people too," Duck said with a shrug, when I asked him why he treated junkies so nice. "Treat them how you want to be treated. Besides," he added, "good customer service is how you beat out the competition."

With Duck killing it in customer relations, I turned my attention to quality control. Crackheads are very particular. A lot of them were complaining about some dope in circulation that had been smuggled into Atlanta in the gas tanks of cars. They said it tasted like fumes and made them feel sick when they smoked it. None of them wanted to buy the shit. But the only way to know for sure if dope was any good was to sample it first. I needed a tester. Butterfly was perfect for the job.

Butterfly was about my age and looked like she might have been pretty before the crack got her. By the time I met her on Baldwin, she wore a ratty blond wig and, as far as I could tell, didn't own a bra. But she was a good worker. Whenever I went to make a buy, I took her with me to test the product. If she gave my dope her crackhead stamp of approval, she'd spread the word faster than if I'd put it on CNN. "Rabbit got that good shit!" she'd tell everybody on the block. Pretty soon, Duck and I were moving half an ounce a night.

IT WAS DUCK'S IDEA for us to expand the business. He thought we should serve twenty-four hours a day. "I'll take nights, you can take days," he said. He put up some money so we could buy more product and we became partners. "We gonna make some *real* money," Duck said.

At first I was skeptical about selling around the clock. I couldn't imagine folks wanting to get high first thing in the morning. But I enrolled my babies in day care and hit the block a few days later, bright and early, at 8 A.M.

It turns out all kinds of people like to start off the day by hittin' the pipe. And not all of them were extra-grimy crackhead zombies who'd been up all night, twitching and fidgeting, with paranoid eyeballs darting every which way. Some of my customers were high-functioning users who'd buy a dime bag from me, showered and dressed and on their way to work. I served janitors, construction workers and a nurse's aid I recognized from Grady Hospital. I even had a mailman for a customer. Mr. Joe would hit me up dressed in his dark blue United States Postal Service uniform. I couldn't figure out how he passed the drug test at his government job. Then one morning, he came by wearing regular clothes.

"You don't deliver mail no more?" I asked.

"Nah," he said. "That shit wasn't for me. But lemme get a dime."

At the liquor house, Granddaddy used to point at the drunks passed out in the living room and tell me, "Baby girl, you see these fucked-up muthafuckas? Don't you *never* drink this shit. Y'hear me? *Never.*" He put the fear of God in me, making me think one sip was going to send me straight to hell. The way he talked, I got the idea that being an addict was a choice. But serving customers everyday on Baldwin, I began to wonder if he was right.

Crack seemed to have a different hold on folks than liquor did. Drunks would sober up and come to their senses in the morning. But once a crackhead got hooked all they did was chase that high. Even if it meant selling everything they owned for a hit: wedding rings, household appliances, their kids' clothes. Anything that had been important didn't matter anymore.

Sometimes I'd feel bad, like when I saw Mr. Joe, who used to look so neat and tidy in his mailman uniform, shuffling down the sidewalk with his TV in his arms, trying to sell his set for a couple of dime sacks of crack.

But other times, like when it was cold and rainy and I was standing on the corner for hours freezing my ass off, I didn't feel bad for anybody but myself. When a crackhead came offering their prized possessions at bargain basement prices, I'd make a deal. I got myself a gold-plated Guess watch with a leather strap, a Samsung VCR player, and an entire set of dinner plates that featured the logo from one of my favorite TV shows. When Duck saw my new dishes he raised his eyebrows. "What the fuck you buy this for?"

"It's from *Dukes of Hazzard*!"

"This ain't from the TV show," he said. "This is a picture of the gotdamn Confederate flag."

DUCK AND I STARTED SERVING on Baldwin in August of 1988. By that winter, we were easily making five or six thousand dollars in profit *a day*. I had so much cash, I asked Duck to hold most of it for safekeeping. He hid it in the back of his closet, at the bottom of his laundry hamper, in the bedroom of the apartment he lived in over in Macon with his girlfriend and their two kids. I guess he figured if someone was desperate enough to dig through his dirty drawers they could keep the money.

But even with everything I gave Duck, I still had more money then I knew what to do with. I dropped a lot of cash on Derrick, buying him dozens of pairs of sneakers and thick herringbone chains. And I spent it on clothes for myself and the kids, filling my apartment with piles of Nautica, Polo, and Tommy Hilfiger gear.

After a while, I couldn't take the clutter, so I moved to a bigger place, a three-bedroom apartment in a complex on Cleveland Avenue. I bought all brand-new furniture from Wolfman Furniture Warehouse, including a white-on-white living room set, a sound system with six-disc CD player, and bedroom suites for the kids. I did Ashley's room in pink and gray and Nikia's in navy blue, red, and white. I did all that and *still* had money to spare. That's when I decided to buy a car.

"But you don't even have a license," Derrick said when I asked him to go with me to the car auction to pick out a ride.

"It doesn't matter," I said. All I really wanted was a place to sit when the weather was bad. I settled for an '83 Chevette hatchback. I handed Derrick six hundred dollars in a wad of small bills. The car sold for four-fifty; Derrick kept the change.

THE MINUTE I BOUGHT THAT CAR, I was itching to drive. I was fifteen, old enough for a learner's permit, but I didn't have a mama or a daddy to teach me. I couldn't call my brothers because Andre and Jeffro were in jail and Dre was busy running his own business, breaking into houses. So instead, I asked Freddy Jack. One of my high-functioning customers, he was the most clean cut and professional looking. By day he worked in a car dealership. At night, he gave me driving lessons. All I had to do was keep him high.

We practiced for hours, circling the block over and over. "Girl, you handling that car like a getaway driver," Freddy Jack said, taking a hit off his pipe and blowing the smoke out his open window. A few nights later, he coached me onto the expressway. "You doing good," he said. "Now *eeeeeeease* into the lane."

In the backseat Ashley bounced up and down. "Mama driving!" she yelled, clapping her hands. "Mama driving!" I gripped the steering wheel with one hand and flicked on the radio with the other. Bobby Brown was singing "My Prerogative," but I had to turn it down when Freddy Jack got too excited doing choreography in his seat.

By two o'clock in the morning, the kids were sound asleep and the roads were clear. Freddy Jack was high as a kite beside me, but I was the one who felt like I was flying.

CHAPTER 16

Mama on the Block

H ey, baby . . . *Heeeey!*" I could hear Mama calling me from
halfway up Baldwin Street. She was rolling toward me in her
wheelchair, holding a big black umbrella against the blazing sun.
From far away, she looked like a handicapped Mary Poppins.

The wheelchair was a new addition to Mama's life, but her
health had been sliding downhill for years. She drank morning,
noon, and night, and smoked a pack of Winstons and a nickel bag of
weed every day. The only thing she had going for her is she wasn't
fat. But somehow she got the diabetes anyway. What put her in a
wheelchair was that one day, drunk as a skunk, she tripped on the
front walk and punched a hole in her foot. When that foot hole got
infected and started oozing pus, Mama tried all her home rem-
edies to heal the wound: hydrogen peroxide, Mercurochrome,
Vaseline. She even put Vicks in it, which burned like a muthafucka.
But nothing helped. Eventually the infection got so bad the doctors
had to chop off her right leg, just above her knee.

At the hospital they gave Mama a fake leg and told her she could
walk again. But she never got the hang of it. The most I ever saw

her take was a couple of shaky steps, clinging to the wall. "Look at me go!" she said one day when I stopped by her place for a visit.

"That's real good," I said. "Keep it up and you'll be doing the robot down the *Soul Train* line in no time."

After she lost that leg, Mama's only transportation was her wheelchair. Lucky for her, Al, her drinking partner, was homeless. He would roll her all over town in exchange for a place to lay his head.

"Hey, Rabbit!" she called again, waving at me as she made her way down the block.

"Hey, Mama, how you been?"

Al pushed her wheelchair to a stop, the two of them grinning at me like we were having a social visit, even though we all knew why they were there. "Baby," Mama began, looking up at me with her hand in front of her face to block the sun from her eyes. "You think you can gimme a little somethin'? I ain't ate nothin' all day. Not one damn thing."

"Mama, you gotta eat everyday."

"Who you tellin'! Just gimme enough for one chicken leg and a biscuit from Church's. And a little extra," she added. "For a couple of bags of reefer, two quarts of beer, and my cigarettes. And somethin' for Al. That's all I need and we'll be straight."

I couldn't believe how things had changed. It was only a few years before that Mama showed me how to diaper my baby with a T-shirt. Now she was coming to me for help. I paid to get her gas turned back on, I helped her with her rent, I made sure she always had enough to eat. I didn't mind giving Mama money. In fact, I liked seeing how it cheered her up. For as long as I could remember, all my mother had to look forward to in life was the thrill she got when she guessed the exact retail price of a dinette set on *The Price Is Right*. But once I started paying her bills, she smiled more and stopped calling me nasty names. That money was like a ray of sunshine in her fucked-up life.

Of course, she still had her moods. One afternoon I went by her place and needed to make a phone call. Mama had one of the welfare phones that President Ronald Reagan gave to extra poor folks. I picked it up, dialed Derrick's number, and Mama hit the roof. "Get the fuck off!" she hollered. "It's *my* muthafucking phone. It's for *emergencies*!"

She was making such a racket, I told Derrick to hold on, then I lifted her out of her wheelchair and laid her down on the bathroom floor. "I'm talking to my baby daddy," I said. "You making too much noise." A few minutes later, I felt a bullet fly right past my head. Mama had crawled out of the bathroom on her stomach, military style, and was shooting her .22 in my direction. But other than that one time, I had to admit Mama had gotten a whole lot easier to be around.

Standing on Baldwin, I reached into my pocket and pulled out a knot of bills. "Those new tennis shoes you got on?" Mama asked as I counted out my money. "They sharp to the bone! And you got your hair done?" she continued. "Look at you lookin' just like that singer El DeBarge!"

Al looked up, confused. "DeBarge?" he repeated. "You talkin' about that light-skinned nigga?"

"Yeah!" Mama said. "He got pretty-ass hair. Don't Rabbit look just like him?"

"Y'all crazy," I said. But I couldn't help smiling. Coming from Mama, this was like telling me I looked like Miss America. She was always good for a compliment when I had a fist full of money. I peeled off a hundred dollars and passed it to her.

"Thank you, baby!" she said with a smile. "You a good girl."

Al took the handles of Mama's wheelchair and started pushing her back up Baldwin. They'd only gone a few feet when Al stopped in his tracks and tilted his head in the direction of a noise coming from down the street. I heard it too: the ice cream truck was pulling up to the curb at the other end of the block. Next to alcohol, weed, and cigarettes, Mama and Al loved them some ice cream.

"Ma!" I called out. "Y'all want something?"

"Ooooooh!" she said. "That sounds good to me!"

It had to be ninety degrees out and there wasn't a lick of shade on the block. I could already taste the icy goodness of a Bomb Pop melting in my mouth. I glanced toward the truck. Little kids, grandmamas, and junkies were already heading that way.

"What y'all want?" I asked Mama, wiping sweat from my forehead with the back of my hand. She was opening her mouth to answer when suddenly the back of the ice cream truck flew open and half a dozen men, dressed head to toe in black military gear and bulletproof vests, poured onto the street. I froze. This wasn't a real ice cream truck. This was a drug bust, and I was holding fifty rocks in a baggie in the front pocket of my jeans.

"Get down!" the cops shouted, pointing their guns at the crowd. "Police! Get on the muthafucking ground, NOW!"

I took a step back toward Mama and Al, keeping one eye trained on the police. They were shoving their knees into folks' backs and pushing their faces to the ground. These weren't regular cops, either; they were the Red Dogs, a special anti-drug unit of the Atlanta Police Department. They'd been raiding traps all over the city for months.

"Red Dog" was supposed to stand for Run Every Drug Dealer Out of Georgia. But on the block we called them the jump-out boys because their favorite tactic was jumping out of undercover vehicles. Their second-favorite tactic was beating the shit out of black folks. From where I stood, I could see a lone Red Dog had broken away from the pack and was coming up the block headed right for us.

"Baby!" Mama whispered. She leaned forward in her wheelchair. "Gimme your dope." I looked down and noticed she'd lifted up the pant leg of her bell-bottoms. Without taking my eyes off the cop, I reached over and slid her my baggie. Mama dropped it inside her fake leg and quickly pulled her pants back down.

"Hey!" the Red Dog yelled as he ran up the street toward us. "Get your muthafucking hands in the air!"

Al and I threw up our hands. The officer headed right to Mama. "What the hell are you doing out here?" he yelled in her face.

"I ain't doing nothin'," she said.

"Are you attempting to purchase narcotics?" he demanded.

Mama let out a gasp and stared at him in exaggerated shock. "Ma'am," the cop continued, "this is an area of known narcotics trafficking. Is that what you're doing out here, buying drugs?"

"*Drugs?*" Mama said, clutching her hand to her chest. "Officer, I don't know anything about no drugs. I don't touch that shit. I'm just out here visiting my baby girl."

The officer turned to eyeball me, then looked back at Mama, frail and thin in her wheelchair. Then back to me. "What are *you* doing here?" he asked, waving his gun in my direction.

"I'm with my mama," I said. "We visiting."

"Visiting who?"

"Visiting each other. She's visiting me. I'm visiting her."

"Is that right?" the officer said, sounding skeptical.

He looked around and his gaze landed on a car parked by the side of the road. "Whose vehicle is that?" he asked, pointing to a souped-up '82 Cadillac Fleetwood. It was sitting on big-ass twenty-two-inch Trues and Vogues tires with spoke rims, and painted pearl white with gold flecks that sparkled in the sunlight.

The car was mine. I'd bought it a few months earlier, right after I turned sixteen and got my license. It cost eight hundred dollars at auction. I dropped five thousand for the paint job and another G for the tires, rims, and custom-fitted Panasonic sound system with a pop-out radio and state-of-the-art CD player with a floor-mounted joystick to control the volume. Nikia's baby-blue diaper bag was on the front passenger seat. Inside the diaper bag was another half ounce of dope, wrapped in a Ziploc. Also in the bag was Derrick's .38 pistol, which he told me to hold on to for protection.

"I never seen that car before," I said, shrugging my shoulders as best as I could with my hands in the air.

When Duck first saw my tricked-out Caddy, he was heated.

"What the hell you doing?" he asked. "Tryna blast the news to every police in Atlanta that you selling dope? Why don't you just put up a gotdamn billboard?" At the time I thought he was overreacting—after all, Duck had taken his drug money and bought himself a double-wide pickup truck, so obviously we had different tastes in automotives—but looking at my flashy ride with fresh eyes, I had to admit, it did kinda scream, "DRUG DEALER!"

The officer leaned over and peered inside the car window. *This is it,* I thought, my heart racing. *If he sees the gun in my diaper bag, or takes a look inside Mama's fake leg, I'm busted for sure.* I held my breath and prayed to God to get me out of this mess. *Please, God,* I begged silently. *Please please please . . .*

Just then, there was a ripple of commotion at the other end of the block. The Red Dogs were loading half a dozen handcuffed men into the back of a police van. I recognized some of them as regular customers, along with a couple of small-time corner boys. They probably didn't have more than thirty rocks between them.

"You niggers are going down!" one of the cops yelled triumphantly. Then he took his baton and smacked one of the handcuffed boys across the back of his knees, knocking him to the ground. "Gotdamn lowlife!" the cop yelled, kicking the kid in the back. "Piece of shit."

The officer who was standing with me and Mama looked up and grinned at the display of excellent police work happening down the block. Then he turned back to us. "All right," he said. "Y'all get out of here. Go on home."

"I told you, my baby's a good girl," Mama called to the cop as he jogged down the block to join the rest of his unit. "You don't *never* need to worry about her!" She turned to me and flashed the biggest smile I'd ever seen. If she'd had some teeth, it would have been perfect. "See," she whispered. "I *got* you, girl!"

Mama saved my ass from the Red Dogs that hot summer afternoon. She was thirty-nine years old; it was the last time I'd ever see her alive.

DRE WAS THE ONE WHO FOUND HER. He went by her place and discovered she'd died in her sleep. Then he came over to my place and told me the news with tears in his eyes. I drove to the hospital to identify her body, and called the funeral home to pick her up. Afterward, I went back to my apartment, sat on my white sofa, and tried to make myself cry.

Mama's dead, I thought. *Dead. Dead and gone. Dead as a doorknob . . .*

I blinked hard. But my eyes stayed dry, which made me feel even worse. What kind of child doesn't cry for their own dead mama? I thought maybe some music might help me get into my emotions, so I put Whitney Houston's "Didn't We Almost Have It All" on repeat, leaned back, and closed my eyes.

"Didn't we almost have it ALLLLLLLLL!" I sang, getting swept up by the beauty of Whitney's voice. Then I caught myself. *This ain't a damn sing-along,* I thought. *You need to get to grieving.*

I tried picturing Mama's face. But the only image that popped into my brain was of Mama throwing her head back and gulping down her gin, which didn't bring tears to my eyes, either.

A memory came to me. I was back in third grade, in Miss Thompson's class, and we were getting ready for the school's annual Black History Month show. Every kid in the show had to dress up like a famous black person we admired. My enemy Mercedes was going as Diana Ross; her homegirl, Porsha, was Aretha. Those two bitches thought they had the best parts, but I knew I was really someone special: I was Corretta Scott King; my granddaddy would have been proud. I made a costume out of Mama's winter coat with the fake fur collar, and a big black pocketbook that Dre stole from the Goodwill. Miss Troup helped me write a speech. "My name is Missus Corretta Scott King," it began. "I am the loving wife of Doctor Martin Luther King Junior." I practiced my lines every afternoon with Miss Troup for a week and then, the day before the show, I stood in the living room and begged Mama to come.

"There's a Black History show at my school," I said, handing her the flyer.

Mama was sitting on her dirty sofa, which was also where she sometimes slept, a tangle of bed sheets beside her. On TV, my favorite McDonald's commercial was playing, the one with the black girls double Dutching and rhyming about Big Macs and Fillet-O-fish. "Shuckin' and jivin'," Mama said, nodding at the set. "You see the way these crackers got our babies dancing for them?"

I glanced at the TV and back at Mama. "So can you come to my show?" I asked.

"Yeah," she said. "Maybe."

The next afternoon I sat on the makeshift stage in the school cafeteria with the rest of my class, sweating under the weight of Mama's ratty coat, my heart pounding from nerves. The place was packed with parents, grandparents, aunts, uncles and the little kids. When a skinny boy named Jovan hit the stage, dressed as James Brown in his mamma's church wig, his daddy jumped to his feet, hollering, "That's MY boy!" When Porsha tried to sing "Respect," her mama and all her aunties held up little cameras, flashing away, not even caring that Porsha messed up, singing "R-E-S-C-P-T." Everybody at Black History had family come out to see them. Everybody except me. Mama never showed.

It occurred to me that maybe it wasn't my fault I wasn't crying for my dead mama. Maybe I wasn't full of grief because Mama hadn't given me the kind of Special Memories I needed to feel sad about her passing. All she ever gave me was a feeling of being cheated out of love. And she *did* try to shoot me.

My mind flashed on my own children. *When I die, those two better bawl their muthafuckin' eyeballs out like they supposed to,* I thought. Just in case, I made a mental note that as soon as Mama's funeral was over I'd take Nikia and Ashley out for a day of fun—maybe we'd all go to McDonald's and the movies—so they'd have something good to cry about when I was dead and gone.

The phone rang. I was so relieved to get a break from the grieving process that I sprinted to the kitchen and picked up after the first ring.

"Hello?" I said.

It was the man from the funeral parlor. "I'm calling to find out when you plan to drop off some clothing for your mother," he said.

"What you mean?"

"A special dress or favorite outfit would be perfectly fine."

"But what she need clothes for?" I was still confused.

"Excuse me?"

"You putting her in a coffin, right? She don't need clothes for that."

"Miss—" He paused and cleared his throat.

"Yeah?"

"Miss, we don't bury people in, uh . . . the *nude*."

I was only sixteen and had never been to a funeral before, so I didn't want to argue. But this made no sense. Why would anyone dress a dead body in a perfectly good outfit just to put it in the ground? I told the funeral man I'd bring something over and hung up the phone. Then I pulled on my sneakers and headed to the mall.

I DIDN'T KNOW WHAT THE HELL I should get Mama to wear. I'd never seen her in anything fancy when she was alive. All she ever wore was jeans and loud-colored shirts that were fifteen years out of style. I stood at the bottom of the escalator at Macy's, scanning the store directory, wondering which floor might have fashions from the early seventies. Then it hit me: Mama was going to sleep forever. I should get her a nice set of pajamas.

I took the escalator to the sleepwear department on the third floor and stepped into a sea of pastel-colored flannel. I walked past racks of furry slippers, oversized nightshirts printed with cartoon characters, and big fluffy robes. Mama usually slept in a man's T-shirt stained with beer, so a matching top-and-bottom pajamas set would be a big step up for her. That's what I was searching for when my eyes landed on the most beautiful outfit I'd ever seen. I reached out my hand to touch the fabric and it felt as silky as a pig's ear. I guess you really do get what you pay for, because the price tag said $195.

"Is this for someone special?" asked the saleslady as she folded my purchase in layers of tissue paper.

"For my mama," I said proudly. "She dead."

THAT SATURDAY, Mama lay in her coffin with a full face of heavy makeup, her Jheri curls glistening with activator. The funeral director had dressed her carefully in the outfit I'd provided. She had on a thigh-high peach-colored satin nightie with a matching floor-length robe trimmed in white lace. The set had come with a matching garter, which I could see had been placed neatly on Mama's thigh, right above her fake leg. I was pleased with my selection. It looked exactly like something Katherine Chancellor would wear on *The Young and the Restless*. Mama always used to say, "I wish I had that bitch's money." Seeing her dressed up in fancy lingerie, I couldn't help but think it looked like Mama finally made it.

As I stood by the casket, Dre stepped up beside me. For a few minutes the two of us stood there in silence, our heads bowed over Mama's lifeless body. Then I heard Dre whisper my name. "Rabbit," he said, "why Mama wearing these ho clothes? She look like she on her way to sell pussy in hell."

The funeral was small. Andre and Jeffro couldn't be there because they were locked up, but Sweetie came with her daughter, LaDontay, and her new baby, Diamond. Uncle Sugar Ray was there too, with Aunt Vanessa and some of her eight kids. Even Mr. John showed up. As the cemetery workers lowered Mama's casket into the ground, Mr. John sobbed like he'd lost his very best friend. He was the only one who shed a tear.

I went back home after the service and changed out of my

black funeral jeans and flats, and into blue jeans and Air Jordans. I threw Derrick's gun and a Ziploc baggie filled with ten-dollar rocks into Nikia's baby-blue diaper bag, put the kids in the backseat of my pearl-white Cadillac that sparkled in the sun, and headed back to the trap.

CHAPTER 17

The Breakup

Duck didn't know it, but I loved him like a brother. Not like one of my real brothers, who were a bunch of petty criminals. I loved Duck like the kind of brother I could look up to, like a role model. Duck had a way of doing things that I'd never seen before. Every time we had to make a business decision—like how much product to buy, or where to get it, or who to buy it from—he'd ask around, gather information, weigh his options, and only then decide what to do. He called it "being strategic." In my family I don't remember anybody having any strategy for anything, ever. Unless you call stealing baby formula from the corner store a strategy. Mostly we just went with our instinct. Like when Miss Betty disrespected Granddaddy in front of his customers, his instinct said, "shoot the bitch in the ass."

Watching the way Duck handled himself, and seeing all the money it was making us, I began to think maybe there was a different way to go through life than what I'd been taught. Everything about Duck—his calm spirit, his business strategizing, the way he could have money in his pocket and not spend it all at once—was completely new to me. Duck was my inspiration. I never said it to his face, but secretly I dreamed of being more like Duck. So even though *I* was the one who introduced *him* to selling drugs, when it came to running our business, I fell back, watched, listened, and followed his lead.

Like most dealers on the west side, Duck and I bought our dope from a dude named Mello, whose specialty was a product called "breakdown." A couple of times a week, Duck and I would drive to Mello's trap, buy five or six thousand dollars' worth of breakdown, take the dope back to Duck's sister-in-law's house, lay it out on her kitchen table, break down the rocks into smaller pieces, and bag it up in dime sacks we'd sell for ten dollars a pop. It was a lot of work, but it was worth it. Mello's dope sold on the street for three times what we paid.

Mello ran a strictly wholesale business, only selling to other dealers. But breakdown was so popular, sometimes when Duck and I pulled up to Mello's trap, he'd be all sold out.

"This is some bullshit," Duck said one afternoon as we drove away from Mello's empty-handed. "Can't make money without inventory. We gotta find a new connect."

Duck asked around, got some recommendations, and set up a meeting with a big-time supplier at the construction site of a strip mall off Old National Highway. The next thing I knew, the two of us were driving down I-285 in Duck's cherry-red double-wide pickup truck, with fifteen thousand dollars in cash, on our way to meet some dude called "The Mexican."

Duck pulled into an empty parking lot, turned off his ignition, and reached into his glove compartment. He pulled out a roll of bills held together with a rubber band and his .38 pistol.

He turned to me: "You good?"

I had Derrick's loaded .38 in the zippered inside pocket of my black leather MCM pocketbook. "Yeah," I said. "Let's do this." Neither one of us had ever shot a gun before. But everybody knew you didn't show up for a drug deal without protection.

Inside the empty strip mall, planks of wood and dusty card-board boxes were piled in the corners. Leaning up against a wall smoking a cigarette was the Mexican, dressed in a black-and-white Raiders starter jacket, with a matching cap pulled low over his eyes.

"You Duck?" he asked when he saw us.

"Yeah," said Duck in a voice several octaves lower than usual. I felt like I was doing a drug deal with Barry White.

The Mexican nodded in my direction. "Who the fuck is *she*?"

Duck answered: "She's with me."

The Mexican looked me up and down with a sneer. Right away I knew I had to show him I wasn't a punk. If we were gonna do business together I needed The Mexican to take me seriously.

"Look here, muthafucka," I said, puffing out my chest and stepping toward him. "You better not be trying to sell us none of that gasoline-tasting bullshit."

The Mexican took a step toward me so the two of us were standing a foot apart, staring deep into each other's eyes. If I'd wanted to, I could have leaned in and kissed him on the mouth.

"Yo, bish," he said, "you better shut your fuckin' mouth before I smoke your ass." He pulled open his jacket to flash the biggest gun I'd ever seen, stuck in the waistband of his jeans. It looked like a sawed-off grenade launcher. I felt the blood rush from my face.

I'd been scared before—like the time I was playing craps in a trap house with a couple of dealers; one dude thought his friend was cheating, so he pulled out his pistol and shot him in the stom-ach, right in front of me—but nothing like this. I didn't know what the hell got The Mexican so mad. But anger was radiating off him like fumes. He reached for his weapon and I froze. *I guess this is it,* I thought. *I'm either gonna get my head blown off or I'm gonna pee myself.*

That's when Duck stepped in. He held up his hands like "be cool," and said to The Mexican in his smooth Barry White bari-tone, "It's all good. She won't say shit else. *I promise.*"

He whispered to me: "I mean it, keep your damn mouth shut."

DUCK DIDN'T SAY A SINGLE WORD the whole way driving back to Baldwin. He just gripped the steering wheel and stared straight ahead. When we got to the block, he pulled up to the curb and turned to me. "You can't be acting like that," he said. "Talking to people all reckless."

"Yeah, but—"

"Ain't no 'buts' about this," he said sharply, cutting me off. "Straight up, I thought we was gonna have to shoot our way out of some shit you stirred up. You keep running your mouth, you gonna get us killed."

I hated when Duck got mad at me. It gave me a feeling in my stomach that matched the look on Beaver Cleaver's face the time he broke his dad's car window playing baseball in the street. It didn't help that Duck was lecturing me like he was my daddy. "What you need to do," he continued, getting out of his truck and slamming the door behind him, "is calm your ass all the way down."

I TRIED TO DO BETTER. For weeks I kept my head down, sold my rock, and made my money. I didn't get into a single argument with anybody. Except Derrick, the day he came by Baldwin asking me for cash. I knew damn well he was spending it on some girl named Tinkerbell, because Stephanie had seen the two of them together. Derrick and I got into it, yelling and screaming in the middle of the road, until he clocked me with a closed fist. A little old lady sweeping her porch saw us and called the cops. But other than fighting with Derrick, I kept a real low profile.

I even kept my music down. I had a banging sound system in my Cadillac, and my *instinct* told me to blast 2 Live Crew loud enough to rattle the windows of every house on the block. But instead I kept it low key with a little Janet Jackson.

"Oh you nasty boys . . ." I sang along under my breath one afternoon, leaning up against the side of my car. The kids were in the backseat. Nikia had a bottle and Ashley was eating the fries from

her Happy Meal. I'd just picked them up from day care and was thinking maybe I would call it an early night. I could go by Derrick's place and the two of us could take the kids to the movies. Nikia was too young to appreciate a good story, but Ashley was three years old. If I covered her face during the scary part, we could see *A Nightmare on Elm Street 5*.

"Y'all want to catch a movie?" I asked the kids, leaning into the back window.

Before they could answer, I heard someone call my name. "Rabbit!"

One of my regulars, Theotris, dressed in a filthy T-shirt and scuffed tennis shoes, was coming down the block, headed my way. *Damn,* I thought. Theotris was always a problem. I didn't know exactly what was wrong with him, but when he was high, sometimes I'd catch him standing in the middle of the block having a heated argument with the mailbox.

"What you looking for?" I asked as he approached me.

Instead of telling me he wanted a dime sack, like I'd been expecting, Theotris walked around to the back of my Cadillac, cleared his throat, and spat on my ride.

It wasn't a regular spit, either; it was like he pulled it from deep down in his navel. Like he'd been saving it up just for me. The ball of grayish-brown mucus hit my back fender with a thud. I watched, stunned, as Theotris, apparently satisfied with the way he'd redecorated my car, hopped up and planted his ass on my trunk, reclining onto the back window like he was chillin' at the beach.

In all the months I'd had been driving my car, nobody—and I mean NOBODY—had so much as laid a pinkie finger on it without my permission. Everybody knew that a ride as flashy as mine was strictly for standing back and admiring. That was the whole damn point.

"The fuck you doing, muthafucka?" I yelled, storming around to face him.

"Just maxin' and relaxin'."

"Hell to the muthafuckin' no you ain't 'maxin' and relaxin' on my muthafuckin' car!"

I ran back to the front passenger side, stuck my hand inside the window, reached into Nikia's baby-blue diaper bag sitting on the front seat, moved the Pampers and the crack to the side, and pulled out Derrick's .38. Then I marched back to Theotris sitting on my trunk. I pointed the pistol right between his legs. "You don't get the fuck off my car," I growled, "I'ma turn your dick into a blooming onion."

Theotris jumped up with a start and I watched with satisfaction as he ran back up the street. "That's right, take your crazy ass on home!" I called after him. "You stupid muthafucka!"

Halfway up the block, he turned and held out his arms like Jesus on the cross. "You better watch out, Rabbit. When I come back you 'bout to get your head blowed off! With the power vested in me, I'ma kill you dead."

Yeah, whatever, man, I thought, rolling my eyes.

WHEN GRANDDADDY AND I USED TO WATCH *Georgia Championship Wrestling* at the liquor house, it wasn't just the fighting I liked, it was also the way those wrestlers would put on a show. Seeing grown men growling and barking at each other like angry dogs was my kind of entertainment. But once I started working on Baldwin, I didn't need a TV to see folks acting the fool. The block served every flavor of crazy, and Theotris, with his spitting and threats, was just the special of the day. *I don't need to take the kids to the movies,* I thought to myself, with a laugh. *With this shit, all I need is some popcorn.*

That's when I felt the bullet fly past my left ear.

Theotris was running down the street, a pistol in his hand, shooting right at me. "I'ma kill you!"

Another bullet zipped past me on the right. I took off running as fast as I could toward the only place that looked safe, Miss June's house. I bolted past the fence, through the yard, and up the steps of her back porch. As I ran, I felt a sharp pain across my chest, but I kept on going. All I wanted was to get inside, away from the bullets that were whipping through the air like firecrackers. As I reached my hand out for Miss June's back door, I could hear folks on the block hollering at each other, "Get down!"

I burst into the kitchen, the door slamming behind me. Duck was already inside. He took one look at me and, for the first time ever, I saw panic in his eyes.

"Rabbit," he said, stepping toward me and pointing at my chest. "You been hit!"

I reached for my chest; the front of my shirt was covered in blood.

"Call 911!" Duck yelled to his mother.

"No!" I cried, suddenly going cold with a realization that struck me harder than a bullet. I could hardly get the words out as I stumbled back toward the door: "My babies . . . they still in the car!"

I grabbed for the doorknob but Duck threw his arms around me, pulling me back. "You can't go out there. You been hit!"

Duck had a good grip on me, but he was no match for the superhuman mama strength I had in that moment. I raised my arm and knocked him to the ground. "I gotta get my muthafuckin' babies!" I yelled, flinging open the back door.

"Hold up!" Duck called after me. "The kids are okay! They're safe."

Butterfly, who'd been on Baldwin when the shooting started, had snatched my children out of the car and run with them to the front of the house. When I turned around and saw her standing in Miss June's kitchen, holding Nikia in her arms and Ashley by the hand, I started to bawl.

I reached out for Ashley; my hands were covered in blood. "Baby, come here," I said. But she wouldn't move. She just stared at me with her eyes as big as dinner plates and her little hands squeezing her cheeks, like she was trying to hold herself together.

THE AMBULANCE CAME AND TOOK ME to Grady Hospital. In the ER, an old white doctor examined me. "She's very lucky," he said, turning to face the three student doctors who were hovering behind him with clipboards in their hands. The doctor moved my hospital gown and pointed to the side of my chest, under my right arm. "The bullet entered here, mid-axilla"—he touched me gently with his index finger—"and exited the areola, here." He lifted my gown all the way up, so my full chest was on display.

I wanted to tell the student doctors if they kept staring at my titties, I was gonna start charging. But I could tell by the way none of them looked me in the eye that they weren't interested in anything I had to say.

"Had the bullet entered on the left," the doctor continued, with his back to me, "it likely would have severed her superior vena cava." His students nodded, solemnly. I hated the way the doctor was talking like I wasn't even there. Like I was invisible.

"Hey!" I called out, interrupting his lecture. "What's that mean? *Vena cava* what?"

This doctor barely turned around to answer me. "It means if the bullet had gone in your left side, we wouldn't be having this conversation," he said.

One of the young doctors, a sister with glasses and her hair in twists, glanced up from her clipboard. She must have seen the look of confusion on my face, because she said to me: "He means we wouldn't be here talking about you, because you'd be dead."

CONSIDERING I ALMOST DIED, I guess I got off easy. All I had was a shot-up nipple. And who needs two good nipples anyway?

The real damage was in my head. For weeks, I kept waking up in the middle of the night hearing the crack of gunfire ringing in my ears. I wanted to talk it over it with Duck, tell him how scared I'd been, but I didn't get a chance. Duck was tired of my bullshit. He'd had enough.

He broke up with me while the two of us were sitting on his mama's porch one evening, watching the night roll in. I don't know how I didn't see it coming. The way he laid it out, it was obvious he had a long-ass list of things he was fed up with: me fighting with Derrick in the middle of the street and bringing the cops to the block; me being a smart-ass with The Mexican and almost getting us killed; and now me getting my nipple blown off and bringing damn near a fleet of paramedics and police officers right inside his mama's house, when I knew good and gotdamn well she was holding our dope in her top dresser drawer. All of it together—plus my loud music and flashy car, my hot head and fast mouth—he was done. "You're bringing too much attention," he said. "You're bringing heat to the block."

He tried to be nice about it. "It's not *you* . . ." he added. "You and me will always be cool. It's just I think it's time we did our own thing. You know, separate. You go your way, I go mine. You understand?"

"Yeah, I'm good," I lied. "We cool."

Sitting beside Duck as the sky grew dark, I realized I'd never told him how much he meant to me, or that I looked up to him. I'd never told him he was the only person in the whole world I trusted or how much I appreciated that he'd never done me wrong.

I wanted to say something to him that night. But I didn't know how.

WHEN WE SPLIT UP, Duck gave me fifty thousand dollars in cash, my half of the profit he'd been holding in the paper bag in the bottom of his dirty clothes hamper. He nodded his head in the direction of Ashby Grove, a few blocks away from Baldwin. "Nobody's holding it down over there," he said. "You could set up your own trap. I bet you'd make yourself some real good money, too."

CHAPTER 18

Aim Higher

You'd be surprised how hard it is to find a safe place to hide a Ziploc baggie full of crack. Before Duck broke up with me, this wasn't a problem. I'd pay Miss June a hundred dollars a day and she'd stash a package in her top dresser drawer, right beside her King James Bible. When I was serving on Baldwin and needed to re-up, all I had to do was run inside real quick and grab a few rocks. I didn't realize how good I had it until I went into business for myself, hustling over on Ashby Grove.

At first I tried hiding my packages in bushes, downspouts, or underneath a rock. But with landlords, little kids, and stray dogs, it seemed like there was always somebody sniffing around. Then I started stashing my dope in the mailbox on the corner. I'd tape it to the top of the mail slot. Or, if I saw a police car easing down the block, I'd just toss my dope down the chute. When the mailman came to open up the box at the start of his shift, he'd hand me my package and I'd pay him fifty dollars.

Sometimes I'd pay him with electronics my customers had traded me for dope. Once I gave him a practically brand-new stolen Panasonic VCR; another time I gave him a four-slice toaster, still in the box. He was pretty happy, for a postal worker.

After I'd been on Ashby for a couple of months, we got a new mailman. This guy was not with the program. The first time he opened up the bottom of the mailbox and found my dope, he looked at me with an expression of disapproval I hadn't seen since Principal Dixon whooped my ass for stealing Mercedes's ham and cheese sandwich back in third grade. "Look," he said, scowling at me. "This is federal government property. You can't be throwing your shit in here. You understand what I'm saying?"

Then I was back to square one, trying to find a safe and convenient place to hide my supply. I didn't have a whole lot of options. In fact, as far as I could tell, I didn't have any. That's how I ended up getting the girls involved. It wasn't what I wanted, but I didn't have a choice.

THE GIRLS—Tata, Tomeeka, Cece, and Little Cee—had all come to live with me a few months before Duck and I split up. My cousin Tata was the first to move in. She was Aunt Vanessa's thirteen-year-old daughter. I'd stopped by their house one day and found Tata sitting on the floor, braiding her little sister's hair. I didn't know Tata could do hair. But she was doing a real nice job. Her little sister had a head full of braids and beads, looking like a tiny Rick James.

"Y'all want to come spend the night?" I asked my cousin. I figured we could watch TV while Tata braided Ashley's hair and saved me a trip to the salon. *"Family Matters* is on."

Of course Tata said yes. Not just because I had a big-screen TV but also because Aunt Vanessa's place was like all seven circles of hell. There were more than a dozen people staying in her little shit box of a house, including Aunt Vanessa's eight kids and four grand-kids, her boyfriend Louie, and my uncle Peewee. The place stank like stale cigarette smoke, spilled Schlitz Malt Liquor, and weed.

Tata packed some clothes in a plastic shopping bag and we bounced. Weeks went by and never once did Tata ask me to take her back to her mama's house. Aunt Vanessa never asked me to bring her daughter home, either. That's how my cousin started living with me.

The other three girls came not long after. Little Cee was nine, Cece ten, and Tomeeka eleven. They were Derrick's sister Darleen's kids. But sometimes Darleen got distracted from her parenting by her love of smoking crack. I went over to her place late one night looking for Derrick and found the three kids home alone, sitting in front of the TV.

"Where's your mama at?" I asked.

"Don't know," Tomeeka said with a shrug, her eyes still glued to the flickering set.

"Y'all had something to eat?"

"Nah."

I took them out to Lilly's Soul Food, then back to my place. That first night they slept on blankets I spread on Ashley's bedroom floor. After a few days, I realized their mama wasn't asking me to bring them back, either. So I bought two sets of bunk beds, and the girls settled in.

With six kids living with me, my place felt like a group home. In the morning I'd get the kids breakfast and make sure all the girls' hair was combed. There were clothes and hair clips, tennis shoes and Pop-Tarts wrappers all over the place. Then I'd drive the babies to day care and drop the older girls at school. In the afternoon they'd meet me on Ashby Grove. They'd hang out on their friends' porches, listening to music and making up dance routines, while I served my customers. When it got late, we'd all pile into my Cadillac and go home.

One night, the girls announced they'd decided to form a professional dance crew, the Ashby Grove Girls. They pushed back the furniture, got in formation, and showed me the choreography they'd been working on; a perfectly synchronized routine featur-

ing the Roger Rabbit, the Wop, and the Cabbage Patch, with Little
Cee dropping into splits as a grand finale. I was so impressed, the
next day I went to the mall and bought four matching extra-large
neon orange and black T-shirts, with coordinating sunglasses and
orange hair scrunchies. Those girls could dance to anything, but
their best routine was to Young MC's "Bust a Move," which they
played fifty times a night.

At thirteen, Tata and Tomeeka were only three years younger
than me, but I was the one holding us down. I kept everybody fed
and bought them all new clothes. I made sure they went to school,
did their homework, and got their hair did every weekend. When
CeeCee decided her personal style would be enhanced by a gold
tooth, I was the one who took her to the dentist and picked up the
hundred and fifty dollar tab.

It was a lot of work taking care of six kids. I'd holler and complain and tell them they were trying my last nerve, but really I
loved every second of it. Having all that noise and commotion in
the house gave me a family feeling. Sometimes I'd lie in bed listening to them giggling in the other room and think about how we
were having the kind of good times I'd only ever seen on TV. Like
on the *Brady Bunch* when all the kids enter a talent show. Only my
girls really were talented. And instead of Alice cooking up a pot
roast in the kitchen, it was me sitting at the kitchen table chopping
up an ounce of crack.

IT WAS SUMMERTIME when I first asked the girls to hold my packages. I figured they were already out on Ashby Grove anyway, listening to the radio and practicing the Running Man. It only made
sense that they should help me out.

"Put these in your drawers," I said one morning, handing each
girl a packet filled with a hundred dime sacks. At first they didn't
know what I meant. "Put them inside your panties," I explained.
"And hold them for me until I tell you." Then it just became part of

our morning routine. Sometimes Tata or Cece would hold down a corner, serving any customers who came through. All they had to do was reach into their underpants to make a quick sale.

Once in a while I got a nagging feeling that maybe I was doing something wrong. Before they'd come to live with me, none of those girls knew a thing about selling dope. They were good girls. They went to school; they didn't talk back or mess with boys. Now here I was, practically their mama, bringing them into the game. It nagged me like a mosquito buzzing in my ear, but mostly I tried not to think about it. I was teaching the girls to survive, just like my mama taught me.

I STOPPED DEALING WITH SUPPLIERS like Mello and the Mexican when I started running my own trap on Ashby Grove. Instead I found myself a new connect, Lamont, who wasn't anything like other dealers I knew. He'd graduated from college, and during the week he worked in a fancy office building downtown. He said for him selling dope was "strictly supplemental income," which he was due on account of the bullshit he endured as a black man in corporate America. "Money-wise, we got to level the playing field," he said.

Lamont wore khaki Dockers, Hush Puppies shoes, and round, wire-rimmed glasses. He read the front section of the newspaper and was one of the only people I'd ever met who actually voted.

Lamont also had a special gift for making me think about things like no one ever did before. One time he asked me how come I had two black eyes. I told him Derrick punched me in the face. Everybody else in my life—Stephanie, Duck, Miss June, even Mama before she'd passed—always said the same thing about Derrick: "He's no damn good." But Lamont shook his head and asked, "Don't you think you deserve better than this?" For weeks I couldn't get that question out of my head.

Lamont wasn't just good at conversation. He also had the

added bonus of running his crack distribution business by delivery. I didn't have to go to some abandoned strip mall with a loaded gun to get my product. Instead, every Saturday Lamont would come by my place with a package and collect his money.

I was sound asleep one morning when I heard him knocking. I jumped out of bed, rinsed my mouth, and rushed to the front door to let him in. I was more excited to see Lamont than usual. I wanted to show him: my brand-new dinette set.

I'd bought it at Wolfman Furniture Warehouse. It had a glass top and a fake cherrywood frame with matching claw-foot chairs. On top of the table I'd carefully placed a blue glass centerpiece bowl and filled it with plastic fruit, just like they did in the display model in the store.

"How you like my decorations?" I asked, waving my hand in front of the set. "It's nice, right?"

Lamont ran his hand along the top of a chair and picked a fake green apple out of the bowl. He palmed it in his hand like a baseball. For a minute I thought he was going to throw it at me. But instead, he put it back in the bowl and sat down. "You like this table, huh?" he asked.

"Yeah," I said. "Why, you think something's wrong with it?"

Lamont leaned back in his chair and glanced around my living room. I watched his eyeballs move from my cream-colored lacquer entertainment console with the big-ass speakers, to my white imitation leather sofa with the lime-green throw pillows, and the floor lamp that bent over the sofa like a rainbow.

Lined up neatly against the baseboard were dozens of pairs of sneakers in practically every size for the six kids in the house. And on the wall, I'd hung a framed poster of a muscular black man embracing a black woman dressed in a flowing white gown. Lamont might not like the dinette set, but at least the rest of the place looked good.

"You want my *honest* opinion?" he asked. "There's nothing *wrong* with this table," he continued, not waiting for a response.

"But every clown with a little dope money and a layaway plan has a set exactly like this one. It's nothing special. It's nice, but it's just *ghetto* nice."

"What's that supposed to mean?" I asked, my face getting hot.

"The thing about money is it gives you choices," he answered, leaning back in his chair. "You can spend it on bullshit or you can invest in quality. Hood niggas are good for dropping bookoo dollars on bullshit, flashy shit that shines bright but doesn't last. That's the ghetto mind-set. You see where I'm coming from?"

I stared at him hard because *no, I did not see where he was coming from*. All I could see was that he was in my house insulting my dinette set.

I'd been doing business with Lamont for months, but he'd never talked to me like this before. Usually, he'd just ask me questions about why I had so many kids in the apartment and how come I didn't have anybody helping me take care of them. One time I told him about living in Granddaddy's liquor house, and the way Mama would fire her pistol in my direction whenever she got mad. "Man," said Lamont, whistling through his teeth. "That's some deep-ass shit. Like *multigenerational* deep-ass shit."

Usually I felt pretty good talking to Lamont. But I didn't appreciate this particular lecture. I crossed my arms and glared at him. He looked at me with raised eyebrows, then threw his head back and laughed.

"C'mon now, don't start tripping," he said. "I'm just trying to school you. As matter of fact, go get dressed and put something nice on. We're going window-shopping. I'm gonna show you exactly what I'm talking about."

I don't know where I was expecting Lamont to take me, but it certainly wasn't where we ended up: a high-end strip mall on Peachtree Road. Lamont pulled into the parking lot, opened the passenger door for me, and led me toward the big glass doors of a store called Haverty Furniture. At first I thought it was some kind of joke. I didn't know anybody who shopped on this side of town. But when I stepped inside the showroom, I almost gasped. It was filled with the most beautiful home decor I'd ever seen. There were five-piece bedroom suites that cost more than a thousand dollars, and dining tables made of chocolate-colored wood. We walked through displays of leather sectionals and tested out the furniture, sinking into a pair of deep armchairs. "This is what I'm talking about," said Lamont, leaning back and putting his feet up on a matching leather footrest. "This is quality. Remember that."

Lamont started taking me on all kinds of field trips. We drove through beautiful Buckhead, along quiet streets lined with Hollywood-style mansions and manicured lawns. We went to luxury car dealerships and test-drove Mercedes. One day Lamont drove north on the I-75, out to a newly built subdivision in Peachtree City, where we toured a model home.

"See that?" Lamont asked, pointing to the kitchen's granite countertops. "That's some high-end shit." He showed me "quality" walk-in closets with built-in shelves, and "quality" crown molding in all the rooms. As Lamont guided me around the place, I noticed a white lady gripping the arm of her husband and pointing our way. She had on a cream-colored pantsuit. I was wearing a black off-the-shoulder T-shirt with a neon-green lightning bolt. I stepped to her and opened my mouth to ask, "The fuck you looking at?" but Lamont pulled me away.

One night we went to dinner at LongHorn Steakhouse and Lamont showed me how to use a pepper mill. I sometimes got the feeling that Lamont was treating me like he was Mr. Miyagi and I was the Karate Kid. Only instead of showing me how to do a crane kick, he was on a mission to teach me about life outside the hood. I asked him about it once, when he dropped me back home: "Why you taking me all these places?"

"Each one teach one," he said, as though that explained everything.

I didn't understand Lamont taking me out and showing me the high life any more than I knew why Miss Troup cleaned my clothes and did my hair, or why Duck had put up with my shit for so long. All I know is that Lamont opened my eyes to a different way of living. "You gotta look outside the ghetto if you want to get ahead," he said. "The hood is nothing but a trap."

MAYBE I DID FEEL A LITTLE BAD about teaching the girls to hustle, because after Lamont started introducing me to his bougie ways, I made it my mission to show the girls everything I learned. I took them to Red Lobster and Myrtle Beach. We sampled all the food at the buffet at Golden Corral and feasted on prime rib at the Sizzler.

One Sunday afternoon, we took a special trip out to Peachtree City so I could show the girls a model home. We wandered from room to room, the kids oohing and ahhhing at the sunken living room with wall-to-wall carpeting and the master bathroom that was big enough to roller-skate in. Afterward, we all piled back into my Cadillac. Before I turned on the ignition, I told the girls to listen up because I had something important to say.

"You see that house with all that beautiful shit?"

"There wasn't no food in the fridge," said Nikia.

I ignored him and kept talking: "One day all of you can live in a house just like that, with a big-ass bathroom and carpets everywhere." They were silent, like maybe they thought I was talking shit. "I'm *for real*," I said. "You can do anything and be anything you want in this life. All you have to do is dream."

I WAS SIXTEEN YEARS OLD and I happier than I'd ever been. I had a family, money coming in, a nice car, and someone in my life who cared enough about me to teach me the difference between bullshit and quality. After all those years of going to school hungry and smelling like dirty Goodwill clothes, finally everything was going good.

All I needed was for Derrick to settle down and stop fucking every girl he met, and my life would be 100 percent perfect. Driving away from the model home in my pearl-white Cadillac, bumping Bell Biv DeVoe on my Panasonic sound system with my babies falling asleep in the seat beside me. Tata, Tomeeka, and Cece singing in the back, I couldn't imagine a better time.

That's how stupid I was back then. It didn't even occur to me that if you do illegal shit all day every day, sooner or later you're gonna get caught.

CHAPTER 19

Locked Up

Everybody knew Officer Harris was out to get me. Blond haired, blue eyed, and a total asshole, he was the beat cop who worked on Ashby Grove. His signature move was slowing down his patrol car and leaning out the window with his fingers curled like hooks, pointing at his eyeballs and then at me. "I got my eyes on you," he'd say. Like that was supposed to scare me.

He'd do this corny-ass move so much that at home the girls and I would point to our eyeballs and back at each other, like, "Gimme the remote. I got my eyes on you." Or "These your dirty tennis shoes on the sofa? I got my eyes on you." Officer Harris was a joke. At least that's what I thought, until the day he decided to take me down.

IT WAS JUNE, a few weeks before school let out for the summer. I was getting the babies ready to drive them to day care and the older girls were heading out to school. Except Tata. "There's no school today," she announced. "I don't have to go."

I rolled my eyes. Tata was going through a stage where she was trying all kinds of bullshit just to see if she could get away with it. "For real," she insisted. "It's a school holiday."

"I guess it's Tata Day," Ashley muttered to herself as she passed me on her way out to the car. Ashley was only four years old, but even she knew Tata was full of shit. I didn't have the energy to argue so I told Tata, "If you not going to school, you coming to Ashby Grove with me."

I didn't drive my Cadillac that morning. I took my new car, a dark-green station wagon that I'd bought at the car auction a few weeks before. It wasn't flashy like my Caddy, but it had more room for the kids. It was my Big Mama car.

I dropped Nikia and Ashley at day care, then drove to Ashby Grove and parked at my usual spot in front of the laundromat. "Get out the car," I told Tata, instructing her to take care of any customers who came by while I walked over to Lilly's Soul Food around the corner to get us something to eat.

I couldn't have been gone more than twenty minutes. I was headed back, carrying a take-out container of biscuits and gravy, when Jerome, one of my regulars, pulled over in his beat-up white Corolla and flagged me down, waving frantically out his car window. "Rabbit, get in the car!" he yelled. "Get in the car!" Jerome was always jittery, but now he was hollering like some body had been shot. "Girl, you better hurry up!"

"What's going on?" I asked, leaning into his car.

"I was just over by your spot and I seen Officer Harris hiding in some bushes."

"What you mean 'hiding'?"

"Rabbit, him and his partner got binoculars like they're doing *surveillance*. Girl, they watching your trap like they finna bust your ass!"

All week long I'd felt like eyeballs were watching my every move. Now I knew why. Officer Harris was a sneaky-ass muthafucka. He was just the type to lurk in the bushes. I wouldn't be surprised if that's what he did on his day off, just for fun.

I opened the car door, slid into the passenger seat, and told Jerome to drive up the road and pull over at the top of Ashby Grove. The street was on a hill and I knew I could get a clear view of what Officer Harris was up to from the top of the block. While Harris was watching my trap, I'd be watching him.

"I feel like that nigga Eddie Murphy," Jerome said as we pulled up to the curb.

"Huh?" I asked, not taking my eyes from the road.

"*Beverly Hills Cop!* You know, like we on a stakeout."

I ignored Jerome and slid low in my seat. Through the dusty front window I could see down the block to the mailbox where I used to stash my dope, the bushes where Harris and his partner were hiding, and my green station wagon parked outside the laundromat.

I'd bought that car because it had more room for the kids, but lately I'd been catching Tata with her boyfriend in the back with the seats laid down, and the two of them hugged up side by side, their faces pressed together.

"You better be using protection," I told her at least once a day.

"Nah, we just kissing," she always said.

Tata was fourteen, the same age I was when I had Ashley. I knew more than anyone how a boy can whisper in your ear and take you places you don't want to go and the next thing you know you got two kids and no way to take care of them. Sitting in Jerome's car, staring down at the street, I made a note to myself that maybe it was time for me to give Tata a "you better not get pregnant" talk. When I was younger, all I had was Sweetie telling me I owed it to Derrick to give him some ass. But Tata was lucky, she had me. I could give her the kind of sex talk I wish I'd had *before* I met Derrick. I started making a mental list of everything I wanted to tell her.

One: Respect yourself

Two: God does not want you to be no ho.

Three: I will beat the stone cold shit out of you if you turn up pregnant.

Just in case, I decided I'd also bring her to the Free Clinic and get her some birth control. After all, I couldn't be with her every second of the day telling her to keep her legs shut.

"Hey," whispered Jerome. "Ain't that him? He's moving!"

I'd been so focused on the talk I was gonna give Tata, I'd taken my eyes off the bushes where Officer Harris had been hiding. Jerome was right. Harris and his partner had left their hiding spot and were running down the block, crouched down low, their hands on their holsters. My heart jumped into my throat. The two of them were running right toward my station wagon. Only, Tata wasn't on the sidewalk where I'd left her. She was inside the car.

HARRIS AND HIS PARTNER must have had it all planned out, because they moved down the block like it was choreographed. Harris took out his baton and rapped on the rear window. In one swift motion, he pulled open the car's back door and he and his partner dragged Tata and her boyfriend out by their legs. I watched my cousin get pushed to the ground and handcuffed. Then Harris crawled into the back of my station wagon and started tearing up the place, searching for drugs.

At the other end of the block, a patrol car pulled up, with its siren blaring. A female cop stepped out. She had a word with Harris's partner, then walked over to Tata, yanked her to standing, and did a pat-down, running her hands up Tata's pant legs, across her back, and over her arms. The officer reached into the front of Tata's jeans and pulled out the baggie filled with fifty rocks I'd given Tata to hold that morning.

"Gotdamn," said Jerome under his breath. "Caught red-handed."

The two of us watched in silence as Officer Harris shoved Tata into the back of his patrol car and drove away.

Tata got popped a few days after my seventeenth birthday. A week later, Officer Harris busted me, too.

I had a good lawyer, the one all the dealers used. I paid him twenty-five hundred dollars in cash, and he got me off with one year probation. But I was so hardheaded, I refused to check in with my probation officer after the first meeting. I didn't like her attitude and the way she talked down to me like I was some kind of idiot. I didn't tell my P.O. I wasn't going to make my next appointment; I just didn't show up. When the cops picked me up a few months later for violating my probation, I got sentenced to a year at Fulton County Jail.

IT'S FUNNY HOW FAST things can change when you don't even see it coming. One minute I was on top of the world, admiring a model home with marble countertops and crown molding in every room, thinking, *I'm gonna live in a place like this.* The next I'm behind bars, eating food that tasted like shit-covered cardboard, and thinking about all the ways I'd fucked up my life.

That was the worst thing about jail. There was nothing to do but think.

I couldn't stop worrying about Tata. She got sent to juvenile detention for a year, but as soon as she got there she found out she was pregnant. She gave birth to her baby boy behind bars, and it was all my fault.

Mostly I stressed about my kids. I'd asked Derrick's friend Slim and his wife Mary, who had three children of their own, to keep my babies while I was away. They were decent folks and I thought my kids would be okay. Nikia sounded fine when Slim put him on the phone. But every time I called the house to speak to Ashley, I could barely get her to talk. "You okay, baby?" I'd ask.

"Yes, ma'am," is all she'd ever say. Her voice was so quiet, it sounded like she was talking to me from the bottom of a well.

FULTON COUNTY JAIL WAS NOISY and hotter than fish grease. We wore orange jumpsuits and shower slippers with our hair sticking up every which way from perm withdrawal. A lot of the inmates were mothers like me who did some dirt—like turning tricks, or holding dope for their boyfriends— to get money to take care of their kids. One girl was coming down off heroin. She spent two weeks puking her guts out. Another girl looked like a walking skeleton, covered in purple sores. "She got the AIDS," my upstairs bunkee, Eva, whispered to me. "It's like they just keeping her here to watch her die."

It was depressing as hell. To make things worse, all anybody did was argue. I once spent an entire day listening to a girl named Jamilah arguing with her cellmate about whether some other chick named Rhonda deserved a beatdown for stealing the Jumbo Honey Bun Jamilah got at the commissary. I didn't have time for petty bullshit and triflin' hos. Instead I found the only two girls at Fulton County Jail who had something positive to talk about. Brenda and Eva became my only jailhouse friends.

Eva slept on the bunk above mine, but I knew her before, from Ashby Grove. She was eighteen, like me, but also a full-time crack addict and part-time prostitute. That's how she got popped: trying to sell an undercover cop a five-dollar blow job. As much as I hated being locked up, to me it looked like jail had done Eva good. As my customer, she'd been skinny and twitchy. Without crack in her system, she had bright eyes, good skin, and a little meat on her bones.

"I gotta be real with you, Rabbit," she said. "I feel like I got a new lease on life." At night we'd lay in our bunks and Eva would sing to me until I fell asleep, Mariah, Janet, Whitney, whatever I requested. She had a beautiful voice. I thought for sure she could have been a professional backup singer if the crack hadn't turned her into a ho instead.

Our friend Brenda was thirty-four, old enough to be our mama. But she wasn't like any mama I'd ever seen; she was classy and educated. When I asked her how she got locked up, she waved her hand like she was shooing away a mosquito and said, "white collar." I didn't know what the hell that was supposed to mean. But Eva asked around and found out Brenda had been running a check-forging scam, using fake IDs and stolen checks to buy luxury items all over Atlanta. "She's a *baaaad* bitch," Eva said, impressed.

A lot of girls thought Brenda was uppity. But I liked the way she carried herself, with her head held high, like she didn't belong behind bars with the rest of us. It was as if Dominique Deveraux from *Dynasty* had come to jail. Brenda had pretty hair she pulled up in a twist, long fingernails, and fake breasts that filled out her orange jumpsuit like two firm grapefruits.

I was so dazzled by Brenda's bougie ways that I followed her around like a lost puppy. The only thing I hated was how she talked like a white girl. Not a regular white girl who works at McDonald's, either. More fancy, like the ones who dress in all black and spray perfume in the cosmetic department at Macy's. When she wanted me to repeat something, Brenda would say, "Pardon me?" It took a while before I figured out that pardon isn't just when somebody lets you out of prison. It also means "huh?"

Brenda, Eva, and I spent all our time together, making each other laugh, complaining about the shitty food and talking about how much we missed our kids. The day I hit rock bottom, they were the ones who pulled me back up.

I WORRIED ABOUT NIKIA while I was locked up, but he was only three years old. As long as he had his collection of Teenage Mutant Ninja Turtles, I knew he'd be straight. It was Ashley who kept me up at night. She'd always been quiet, and she had the worst case of nerves I'd ever seen in a little kid. Before I got locked up, I used to wake up in the middle of the night and find her standing in my bedroom clutching her nightgown, telling me she had a dream that the two of us were standing on Ashby Grove and somebody blew my head off. The only way to get her to go back to sleep was to let her crawl into bed beside me.

The summer before I got put in jail, I'd been planning for Ashley's first day at kindergarten. I'd picked out the perfect outfit at the mall: a white collared Polo shirt with matching Polo jeans and a little kid Falcons starter jacket. When I got sentenced, I told Derrick to make sure to bring the clothes, which were folded up neatly on the top shelf of my closet, over to Slim's place. I knew from experience that next to jail the place with the highest concentration of trifling bitches was elementary school. Ashley was so timid, I was worried the girls at her new school would sense her weakness and make fun of her the way kids had teased me. That's why she needed to look fresh. Nobody makes fun of the best-dressed girl in the class.

On Ashley's first day of kindergarten, I was so nervous I asked my friend Melodie, who lived near the school to do a drive-by and check on how Ashley looked. I leaned my head against the cinder block wall of the jail's dayroom and dialed Melodie's number. "Girl," I said, when I heard her voice. "You gotta tell me the truth."

I could hear Melodie take a breath and slowly exhale. She'd seen Ashley in the schoolyard, she said. But instead of the good-looking outfit I'd picked out, Slim and his wife had let my baby leave the house in high-water jeans, scuffed tennis shoes, and uncombed hair. "Real talk," said Melodie. "Your little girl looked raggedy as hell."

That night I tried to sleep, but I couldn't get the pitiful image of Ashley looking like nobody gave a damn out of my head. Knowing that my baby was out in the world without me felt like a knife through my heart. I felt so guilty I could hardly breathe. I pulled my blanket over my head and cried my eyeballs out.

The next morning wasn't any better. I ate my breakfast, shuffled past the girls watching *The Price Is Right* in the dayroom, then went back to my cell and crawled into my bunk. Ever since I started hustling, I was able to fix all my problems with dope money or stolen electronics. But this was different; I felt helpless. It was the lowest I'd ever been. After days of me barely talking, Eva and Brenda decided I needed an intervention.

"Girlfriend, you gotta pull yourself together," said Eva, standing in front of me with her hands on her hips.

"She's right," added Brenda, who was sitting on the edge of my bunk. "You can't let this incarceration drag you down. You've got to rise above. What I find helpful is creative visualization. It's a very powerful tool. Are you familiar with Shakti Gawain? Life changing."

Eva and I shot each other a look. Brenda was cool, but sometimes we didn't know what the hell she was talking about. *Shakti Gawain?* That shit didn't even sound like English.

"Yeah, I don't know about all that," said Eva. "But I *do* know what you need is a little Whitney in your life!" She closed her eyes, threw her head back, snapped her fingers to keep time, and started to blow, *"I believe the children are our future . . ."*

"That's right!" exclaimed Brenda, jumping up and pulling me off my bunk. She threw her arm across my shoulder. *"Teach them well and let them lead the way!"*

I thought nothing could lift my mood. But when I closed my eyes and let the music wash over me, for a moment, I forgot how shitty I felt. Eva, Brenda, and the power of Whitney Houston reminded me that I needed to stay positive because children are the future and learning to love myself is . . . *the greatest love of alllllllll.*

The three of us belted out Whitney at the top of our lungs until a guard came by and told us all to shut the fuck up.

CHAPTER 20

Hood Wisdom

Getting locked up is good for two things, having same-sex relations in the shower room and making Life Goals. Only one of those things interested me.

At Fulton County Jail, Brenda, Eva, and I spent a lot of time sitting in my cell planning for the better lives we were gonna have once we got out. Eva was going to make amends with her family and try to get back some of the stuff she stole from them when she was getting high. Brenda was going to marry her rich white boyfriend, who was old as hell and probably gonna die soon, leaving Brenda all his money. Those girls had ambition. Hanging out with them, I started to dream big, too.

The plan was for me to get my GED. Brenda put the idea in my head. She said, "If you want a real job, you need an education." I wanted to ask her how come with her two college degrees she didn't have a real job, but I didn't want to interrupt the pep talk she was giving me.

"You're a smart cookie, Rabbit," said Brenda. "And you obviously have a head for business. All you need is focus and determination. After you get your GED you can apply to technical college, or maybe even business school."

"Nah," I said, laughing at the thought. "That's not for me."

"Why not? Most of the guys I date went to business school. Trust me. Those places are filled with nothing *but* fools!" She looked at me. "Nothing personal, sis. I'm just saying, don't sell yourself short."

I'd never thought of getting a GED before. But Brenda said I should set an example for my kids: "If you want them to graduate, show them you can do it, too."

"She's right," added Eva. "The apple don't fall far from the tree. You never heard that saying before? You the tree; your kids is the apples. You don't want them rolling away, willy-nilly, ending up in some gutter somewhere, do you? Nah, you want them to look up to you and say, 'One day I'ma be a tree like my mama.'"

"You stupid," I said, laughing.

"Be a tree, girl! Reach for the sky!"

"You could manage a store," suggested Brenda. "Do bookkeeping, become an executive assistant . . ." She ticked off job possibilities on her fingers like a grocery list. "Work in a doctor's office . . ."

"Be salesperson," added Eva.

"*Definitely* pharmaceuticals," said Brenda.

The more we talked, the more my vision came into focus, until I could actually picture myself, clear as day, dressed for work in a neatly pressed blouse with a little metal name tag pinned to the front. I was gonna get my shit right.

OTHER THAN PLANNING MY LIFE GOALS IN JAIL, I only had two other activities to fill my days, worrying about my kids, and lying in my bunk thinking about Derrick. Except for Nikia and Ashley, he was the only person I really missed. In my heart, I knew it didn't make sense. The days of Derrick making me feel good had long gone. Mostly, all he did was take my money, beat my ass, cheat on me, and tell me I was ugly. One time he stomped me in front of Ashley so bad she ran to the phone and called 911 screaming, "Daddy killing Mama!" But as bad as he treated me, Derrick had a hold on me. It was like he had me brainwashed. When he told me nobody else would ever love me, I believed him.

In jail I only I remembered the good times. Like before I got pregnant, when he took me roller-skating and held my hand; or the time we took the kids to the drive-in to see *Halloween 5: The Revenge of Michael Myers;* or the day we all went to Six Flags and Nikia rode the kiddie roller coaster and threw up cotton candy while Derrick and I laughed our asses off. I played the scenes over and over in my head until all I could recall of life with Derrick was nonstop laughs.

I WAS RELEASED FROM FULTON COUNTY JAIL on a cold day in January 1992. I'd served eight months and gotten out early because of overcrowding. Derrick came to get me. The first thing I wanted was for him to drive me to Slim's house so I could pick up the kids. As we headed down Marietta Boulevard, I told him things were going to be different. I was turning my life around, I said. "You know, I was thinking like maybe I could get my GED and get some kind of job."

When he heard this, Derrick swiveled his head and looked at me like I'd just sprouted wings and told him I was flying to the moon. Then he bust out laughing. "A *job*?" he said. "Girl, you ain't getting no job! You can't do nothing but hustle. Besides," he added, "you got the kids to take care of. What you need to do is get your ass back on the block and start making that money."

I took a deep breath and stared out the window. I thought about Brenda's list of employment opportunities: *secretary, home health-care aide, paralegal, dental hygienist* . . . She said I could do anything I set my mind to. But Derrick was talking to me like I was a fool. "G fuckin' E D my ass!" he laughed. "Who you think you is, Claire Huxtable? Ha!"

Suddenly, I felt so stupid. What was I thinking letting Brenda and Eva fill my head with crazy ideas? Dreaming big was fine when I was locked up. But it didn't make any sense in the real world. Derrick was right, I had kids to feed and rent to pay. That's the power he had over me. He could make me doubt myself and change my mind like nobody else.

The next week, I borrowed a thousand dollars from Duck, bought myself some product, and within the week I was right back where I started, hustling on Ashby Grove.

THE BLOCK LOOKED EXACTLY THE SAME. The only difference was the laundromat had a new sign out front. Painted on the glass window, it said HOOD WASH & DRY. I thought it was messed up that somebody had labeled the place "hood."

"You think white people have a laundry that says 'Cracker Wash and Dry'?" I asked Jerome, staring at the sign. He laughed and told me that some guy named Hood had bought the place. "Girl, that's why it's called Hood," he said. "Not because it's *in* the hood."

Hubert Hood was a middle-aged man with a potbelly and rectangular reading glasses that sat on the end of his nose. I met him when he came out of his office at the back of the store one afternoon and found me beating the shit out of Duck's cousin Reggie with a giant jug of Clorox bleach.

I'd been in the laundry minding my own business, when Reggie walked in and started hollering at his girlfriend, Tanisha, who was using the washer next to mine.

"Bitch! Who told you to leave the muthafucking house?" he yelled. Then he grabbed Tanisha by the throat, shoved her up against a dryer, and started punching her in the face.

I never liked Reggie anyway, and I definitely wasn't going stand there and let him whoop Tanisha's ass, so I grabbed the bottle of Clorox off the counter and slammed it into the side of his head.

Reggie ignored me and kept swinging on his girl, so I twisted the cap off the bleach and dashed it in his face. Reggie fell to the floor, hollering about his eyeballs being on fire. That's when Hood came running out of his office with a pistol in his hand.

"Whoa!" he yelled. "What the hell's going on out here?" Hood grabbed Reggie by the collar, dragged him over to the sink, and shoved his head under running water. He turned to me: "Girl, you trying to blind this boy?"

"Hell, yeah!" I said. "He was beating on my friend."

Hood looked from me, to Tanisha, then to Reggie, who was soaking wet, with his Jheri curl plastered to the side of his face.

"Well," said Hood. "I can't argue with that."

THE NEXT TIME I HAD TO DO LAUNDRY, I brought my kids. They sat in plastic lawn chairs Hood had set up against the wall. Ashley kept herself busy reading a book her teacher had given her about a bunch of talking bears, while Nikia chewed on the head of a Teenage Mutant Ninja Turtle.

Hood came out of his office and nodded hello. Then he noticed my kids. I could see him giving them the once over. Nikia had a fresh haircut and was wearing a pair of snow-white Jordans, creased Levis, and a crisp navy-blue Polo. Ashley was dressed exactly the same, except her Polo was red and her hair was braided into four neat plaits, each one finished off with a red clip in the shape of a bow. "Those your kids?" Hood asked.

"Yeah."

"How old are they?"

"My girl's five and my son's coming up on four."

"Nice looking. Quiet, too."

"Say hello to Mr. Hood," I told the kids.

"Hello, Mr. Hood," they said together.

"Real mannerly," he muttered to himself, walking back into his office. "I like that."

After that, whenever Hood would see me—which was pretty much every day, since I was selling crack right in front of his laundromat—he'd ask, "How your kids doing?"

He came outside one chilly afternoon while I was leaning up against my car hustling and the kids were in the backseat. Ashley was doing a worksheet on top of her winter jacket, which she'd folded up on top of her book bag to make a desk in her lap. Nikia was pretending to feed his G.I. Joe french fries from his McDonald's Happy Meal.

"How long you gonna keep these kids waiting in the car?" Hood asked.

"I don't know." I shrugged. "Maybe two, three more hours."

"Why don't you send them inside?" Hood suggested. "It's cold. And I don't like the way all these knuckleheads out here acting the fool when the sun goes down. It's not safe." I took the kids into the laundry and told them to sit on the lawn chairs. But Hood motioned for me to bring them inside his office. "I got a TV in there," he said.

When I went to check on them a little while later, Ashley was curled up in an armchair in the corner of Hood's tiny office, reading her storybook out loud, while Hood sat at his small metal desk with Nikia, teaching my boy to count the coins Hood used to make change for his customers.

Hood became my kids' regular after-school activity. They'd be with him for hours, doing their homework or watching TV, while I hustled. Sometimes he'd stay open later than usual, until nine or ten o'clock at night, just to wait on me.

Hood was like a granddaddy to my kids. And sometimes when they were at school, I'd slip inside the laundromat to spend time with him myself. "Hey Rabbit," he said when he saw me standing in his office doorway one afternoon. "C'mon in." He had a stack of bills spread on his desk in front of him, a little calculator, and a notebook. "I'm just taking care of my finances," he said, picking up a bill. He peered at it over the top of his reading glasses, made a note in his ledger book, then he wrote a check and slipped it inside the envelope.

"You sure got a lot of bills," I said after he'd stuffed more envelopes than most folks had utilities.

"Credit cards," he explained. He reached into his back pocket, took out his wallet, and pulled out a Visa card. "I got a twenty-thousand-dollar limit on this one. I don't touch it, though."

"What's it for then?'

"Emergencies."

Hood must have seen the confusion on my face, because he went on to detail exactly how his emergency plan worked: "Say if I need a house repair, or my car breaks down, or my wife takes ill, I got this high-limit card to cover me. You gotta hope for the best but plan for the worst. This credit card is in case the worst happens. It's my safety net." I'd never heard anything like this before. When I was growing up, there was never a plan for those days when the food ran out or the electricity got turned off. Life was a string of one emergency after another, with no kind of net to break the fall.

"How do you get one of those cards?" I asked Hood.

"You gotta have good credit," he said. "You can't get anywhere in this life without good credit."

EVERY NIGHT AT 6 P.M., I'd get the kids some dinner at Lilly's Soul Food and the four of us would sit in Hood's office while he watched the evening news. I don't know why Hood even bothered, all it did was give him bad nerves. Hood got worked up about everything: the recession, Rodney King, Anita Hill, and Mike Tyson going to jail. "Back in my day," he said one evening, "we had cats to look up to, like Jackie Robinson and Dr. Martin Luther King. Now it's like we on a highway to hell." He added, "At least we have Bill Cosby."

As much as Hood worried, nothing raised his blood pressure more than fretting about me and my kids. One night when I was working late and Hood had Ashley and Nikia in his office, I noticed he kept coming to the window and looking out onto the street. It took me a while to realize he was checking up on me. Hood was afraid I was gonna get hit by a stray bullet in a drive by, or that some other dealer was gonna stick me up. He said he wished I would quit hustling and find something safer. But the only idea he ever came up with was winning the lottery. "Imagine you hit the jackpot," he said. "You win the lotto, you could get off these streets for good."

The way Hood worried about me, I probably should have known better than to show up at the laundry with a thick white hospital bandage wrapped around my head. But I wasn't used to having somebody care the way he did, so I didn't think it through. When Hood saw me standing at his office door one morning looking like I just came home from war, he clutched his chest like he was having a heart attack: "What the hell happened to you?"

"Nothing," I lied. "I'm fine."

Hood came around the side of his desk to help me into a chair. He took a seat across from me and held my hands in his. "Who did this to you?" he asked softly. I felt so bad about the way he was looking at me—like I was a little bird that fell out of a nest—I could barely bring myself to tell him the truth. I leaned back, closed my eyes, and took a deep breath.

The day before, Derrick had come by my apartment with a giant hickey on his neck. It wasn't the first time, either. Derrick *stayed* cheating on me. It was the number one thing we argued about. "Who the hell been sucking on your neck?" I demanded. "What bitch you fucking now?"

I guess Derrick didn't have a good comeback, because instead of answering he picked his loaded gun off the kitchen table and smacked me across the face. That's when the gun went off.

Derrick said it was an accident. "The gun made a mistake," he insisted. But he took off before the ambulance arrived, leaving me bleeding on the kitchen floor. I guess he didn't want to explain to the authorities that his gun had a mind of its own.

Hood knew Derrick and I fought like cats and dogs. We'd yell at each other in the street and sometimes Derrick would follow me into the laundry. Hood ran him out of the place plenty of times hollering, "You better leave up outta here before I put my foot up your ass."

Hood hated Derrick. And seeing me that day with my head in bandages and a bullet graze on my skull was too much for him to take. He covered his face with his hands. "Rabbit," he said, his voice cracking, "you know I care about you and those kids like you're my own. Every day I pray you find a good man. But God as my witness, you need to quit that piece of shit Derrick or you're never gonna give nobody else a chance. That's how it works, Rabbit. You gotta let go of the bad to make way for the good."

I TOLD MYSELF HOOD DIDN'T UNDERSTAND my love for Derrick. For weeks, I tried to push his words away. But they kept repeating, like a bad chili dog: *You need to leave that piece of shit.* Hood's words popped into my head while I was lying in bed and while I was hustling on the corner. I thought about them while I was giving the kids breakfast and when I was putting them to sleep. I couldn't shake what Hood had said because deep down I knew he was right.

I could make a list as long as the Mississippi River of all the ways Derrick had done me wrong: he beat me with a roller skate, he kept getting other girls pregnant, and now, on top of everything else, he'd accidentally on purpose shot me in the head.

One night Derrick and I had been out driving and had run out of gas. It was storming with thunder and lighting, and we were blocks away from a filling station. Derrick refused to get out of the car; instead he made me go get the gas. An old man, seeing me walking with a five-gallon gas can in the rain, picked me up and gave me a ride. When he saw Derrick sitting in the passenger seat, sucking his thumb, the old man turned to me: "This young nigga made *you* go get the gas?"

"Yes, sir."

"With all this rain coming down?"

"Yes, sir."

"Girl, that ain't *no* kind of man." Even a complete stranger could see Derrick was a piece of shit.

I'd always thought it was love that made me stay with Derrick, but I couldn't name a single thing I loved about him. And I always said we were a family, but we weren't any more together than the day he showed up at the hospital to meet our newborn baby with another girl by his side.

Staring up at the ceiling one night, I kept asking myself, *Why?* Then it hit me. It wasn't love that kept me hanging on; it was fear. I wanted a *Leave It to Beaver*–style family, but I was a teenage mama with two kids, no education, and a blown-off nipple. I couldn't imagine anybody wanting to make a family with somebody like me. I was afraid Derrick was my only chance. But maybe I was wrong.

I climbed out of bed and knelt on the floor, clasping my hands in front of me. "Dear Heavenly Father," I whispered. "I know I been asking you to get Derrick to act right and stop beating and cheating on me so we could be a family. Well, Lord, I changed my mind. Fuck that nigga. Take away my love for that lying piece of shit. Don't change him. Please, God, change my heart."

CHAPTER 21

Mr. Nice Guy

I guess it's just a fact of life that when you get out of a relationship, even a shitty one with a no-good lying piece of shit, you need some time to heal. I'd been with Derrick for seven years. He was the only man I'd ever known. When I stopped messing with him and finally let go of the idea that we'd ever be a happy TV family, it felt like more than a breakup; it was like I'd cut off my arm.

During the day, I tried to keep busy making money; at night I'd sit in front of the TV and try not to cry. I didn't want to admit I felt so lonely without him, so when anybody asked how I was doing, I'd just say, "Fuck that nigga, I'm good!"

"Oh really?" my friend Niecy asked skeptically, when she came over one night and found me laid out on the sofa watching *The Golden Girls*.

"Yeah," I said. "Staying positive, doing *me*."

Niecy was Dre's baby mama. She'd shown up to my house dressed in hot pants and a black leather vest, looking like the missing member of Xscape, and ready for a night out. She pointed to my outfit: "Honey, just look at you! Rabbit, you a hot-ass mess."

I was dressed in a faded red and green extra large nightshirt with a picture of Santa Claus's happy face and the words HO HO HO on the front. "What are you talking about?" I protested. "These are my *relaxing* clothes. I'm relaxing."

"Girl, please," Niecy said, rolling her eyes. "It's July and you dressed like it's Christmas at the crack house. Stop playing and get your ass up. Chile, you need to get out of the house and have some fun." They were having a lip-syncing contest at Harlem Nights, Niecy said. We were going and she wasn't taking no for an answer. She dragged me to my bedroom, pulled an outfit from my closet—a cream-colored version of the exact same outfit she had on—plugged in my flat iron, pressed my weave, gelled my edges, and sprayed my hair with firm hold Spritz. Then she stood back and gave me the once-over. "Yeah, girl, you gonna find you a new man tonight!"

THE CLUB WAS PACKED when we got there. At the front of the room, a contestant was killing it onstage, lip-syncing Diana Ross and Lionel Richie's "Endless Love," in a half-man, half-woman costume. Every few bars, he'd pivot so we'd see either the long-haired Diana in a bedazzled gown or a white-suited Lionel with a 'fro.

Someone in the crowd yelled, *"Saaaaang,* bitch!" as Niecy and I made our way to a table near the stage where some of her friends were already sitting. I took a seat beside a big-boned brother with a friendly smile and low fade. He had on a blue button-down shirt, looking just like Carlton from *The Fresh Prince of Bel-Air.*

"I'm Michael," he said, leaning over and shaking my hand. "And you are?"

"Rabbit."

"Rabbit?" he repeated, looking confused. "That's the name your mama gave you?"

"No. My mama named me Patricia, but nobody calls me that."

He leaned back in his chair and looked at me through squinted eyes.

"Well, you don't look like a Rabbit to me," he said. "You're too pretty to be named after an animal." I could feel my face getting hot. This was way better than sitting at home watching *The Golden Girls*. Michael asked if he could call me Pat and I nodded yes, thinking to myself, *As long as you keep sweet-talking me, you can call me whatever the hell you want.*

Onstage, a three-hundred-pound Mariah Carey was adjusting her half shirt, lip syncing, *"You got me feeling e-mo-tion ..."* When Michael turned his eyes to the show, I gave him the once-over. He definitely wasn't my type. I liked the roughneck Jodeci look, and Michael was so clean-cut, he looked more like the fifth member of Boys II Men. Still, there was something about the way he smiled at me and called me pretty that got me interested. "You live around here?" I asked, leaning forward. I could feel my one good boob pressing against his forearm.

"I been in the military," he said. "I just got back from Desert Storm." He took out a pen and drew me a map on the back of a napkin, pointing to a gulf. "That's where I was deployed, in Kuwait." I knew about the war from watching the evening news with Hood on the little TV in his office. But I'd never met anybody who'd been there.

"What's it like over there?" I asked.

"Hotter than hell," Michael said, over the music. "But at nighttime, when it's quiet, the desert is real pretty. There's nothing but sand all around, and when you look up you can see every single star in the sky." Michael was leaning in so close I could feel his breath on my neck and smell his soapy scent. He wasn't the kind of guy I went for. But I couldn't help but picture him in his army fatigues standing in the middle of a war zone admiring the stars in the sky.

MAYBE I SHOULD HAVE TOLD Michael I had kids before I invited him over. But I was so excited when he called me a few days after we met at the club that I just blurted out: "You want to come over and watch TV?" Half an hour later, he was knocking at my door.

The first thing he did when he came inside was ask me why I had so many pairs of sneakers, which he noticed lined up against the wall. "They're not all mine," I said, explaining that some of them belonged to my nieces Cece and Little Cee, who still stayed with me from time to time. "And the little-kid Nikes belong to my kids," I added. "My son is five and my daughter is almost seven. She's real smart, her teacher just put her in accelerated reading."

He stared at me hard as we sat down in the living room. "How old are you, anyway?" I told him I was twenty and could see him doing the math in his head. But all he said was "Okay."

"You have any kids?" I asked.

"Nah."

"Why not? You look like the family type."

"I don't know," he said with a shrug. "I guess I always wanted to wait, settle down, and do it right. You know," he nodded toward the TV, "like that." On the screen, Aunt Viv and Uncle Phil were sipping tea in the living room of their Bel-Air mansion, while The Fresh Prince made fun of Carlton's dancing.

"Oh, you mean you wanna be rich before you have kids?"

"No, I just meant married and settled down."

THAT NIGHT MICHAEL AND I talked for hours. He asked me about my mama and I told him how she'd take us to a different church every Sunday to get baptized so we could eat. Michael kept shaking his head in disbelief. "Golly," he said. "I never heard anything like that before."

Michael had grown up with his two parents in Cascade Heights. The neighborhood was only fifteen minutes away from where Duck and I sold dope on Baldwin Street, but it might as well been a different planet. Around that way, you never saw women walking out the house in a roller set and their bedroom slippers, and nobody parked their broke-down hoopties in the yard. Michael didn't know shit about caseworkers, eviction notices, or eating ketchup sandwiches for dinner.

"So you never had your lights cut off?" I asked him.

"Well," he said, trying to remember, "one time there was a storm that knocked out the electricity for a couple of hours. I had to use a flashlight."

I don't know if he was trying to be funny, but that shit cracked me up. I guess Michael liked talking to me, too, because the next day he came back to hang out, and again the day after.

Then one day I needed a favor: my car wouldn't start and I wanted Michael to give me a ride home. I called him from the pay phone and asked him if he could come get me.

"Sure," he said. "Where you at?"

"Working," I told him, and gave him directions to Ashby Grove.

Michael worked at the Simmons mattress factory. When he told me about his job, he was real proud of his "benefits," and his 401(k) savings plan. I told him nobody in my family had a bank account. "Then where do you keep your money?" he asked.

"My granddaddy put his cash in a gym sock and hung it down by his balls," I told him. "He didn't get robbed once." Michael looked totally confused everytime I talked about my family. I didn't think he could handle any more tales from the hood, so when he asked me what I did for a job, I kept it breezy. "Part-time entrepreneur," I said, with a wave of my hand.

I was standing halfway up the block, near the laundromat, when Michael pulled up to the curb across the road in his grey Nissan Maxima. I started heading his way, but before I reached him, JaMarcus, one of my regulars, ran over and stuck his head in the passenger-side window of Michael's car. JaMarcus had cracked lips and matted hair and stank like stale piss. He was holding a pair of brand-new Air Force 1's.

"Yo, my man!" JaMarcus said, shoving the sneakers in Michael's face so he could get a good look. "Twenty dollars! If you don't like these, I got some of them Dennis Rodman joints. You know, with the pump. *Pump up the jam, pump it up a little more . . . !* I can get them right quick. Those going for thirty. But you can have 'em for twenty-five. What size you is, brotha?"

"Nigga, get to stepping before I stomp a mudhole in your ass," I called to JaMarcus as I crossed the road. "He don't want none of that stolen shit."

JaMarcus swiveled to face me, with his hands up, like he was surrendering to the cops. "No doubt," he said. "My bad!"

"But what 'bout you, Rabbit?" he continued. "Girl, you need new kicks? A Guess watch? How about a brand-new coffee pot?"

I ignored JaMarcus, opened the passenger-side door, and slid into the seat. Michael turned to me, his hands gripping the steering wheel, his eyes wide. "What the hell is going on over here?" he asked.

I looked out onto the street. JaMarcus was standing in the middle of the road, cradling his stolen sneakers in his arms. Butterfly's skinny ass was strolling the block trying to make some money, in Day-Glo orange bicycle shorts. At the end of the road a middle-aged couple, both of them higher than kites, were hollering at each other, ignoring the baby boy who was crying his eyes out and clinging to the woman's legs. There were dope boys, crack fiends, and prostitutes all up and down the block doing deals, hanging out, and trying to get high. I didn't know what Michael was talking about. The street looked fine to me.

"What you mean?" I asked.

"All these crazy-looking people out here . . ." he said, shaking his head. "I been living in Atlanta my whole life. I've never seen anything like this street before." He paused, like he was taking it all in. Then he turned to me: "What are you doing out here, anyway? I thought you wanted me to pick you up at your job."

"Yeah. This is where I work."

Michael twisted his neck trying to see up and down the street. "Where do you work?" he asked. "In the laundromat?"

"Nah," I said, laughing. "I work in *front* of the laundry."

"Doing what?"

"Hustling."

He looked at me, confused.

"Michael," I said, taking a breath, "I sell drugs." Most people I knew did drugs, sold drugs, or lived off drug-dealer money. But Michael was different. I didn't know how he would take the news. I held my breath and waited for him to respond.

For a couple of seconds he just stared at me, his eyebrows in knots, not saying a word. Finally he opened his mouth. "You pulling my leg?"

"No," I answered. "This is my trap. I run this whole block."

He let out a long whistle. "Wow," he said, shaking his head. Then he started the car and pulled away from the curb. We drove for a few blocks in silence, with Michael staring straight at the road and me wondering what the hell he was thinking. Finally, I couldn't take it anymore. "You think I'm a bad person?" I asked.

"It's not that," he said. "It's just . . ." He trailed off.

"What?"

"I guess I never thought you'd be doing something like this."

He went on to tell me about his cousin. The boy had been fine when Michael joined the army. But by the time Michael came back home, his cousin was hooked on crack. "He started stealing from his mama, and it just broke my auntie's heart," he said. "So you know . . ." He trailed off again.

The sad look on his face while he talked about his cousin made me think for sure he wasn't gonna like me anymore. *And why would he?* I thought to myself. I wasn't anything special. I sold drugs. He could probably get any girl he wanted. Like one of those bougie girls who worked the makeup counter at Macy's, or even a dental hygienist. But a few days later he showed up at my door. "You came back," I said, letting him in. "I didn't think you would."

"I don't know," he said with a shrug. "I guess I like you."

WE'D BEEN DATING ALMOST FOUR MONTHS when Michael decided he wanted to introduce me to his mama. "What if she don't like me?" I asked, lying in bed beside him. I imagined a church lady who'd stare at me with her church-lady eyeballs thinking about how I'm not good enough for her boy.

"Of course she's gonna like you," he said, laughing.

After a while he added, "Maybe don't say your name is Rabbit."

Michael took us to Red Lobster for dinner. I was so nervous, I couldn't think of a thing to say, so I just put my head down and ate in silence. When the bill came Michael pulled a Visa card out of his wallet. I recognized it at as the same card Hood had shown me that day we'd talked in his office, the one with the high limit. "You can't get anywhere in this life without good credit," I said suddenly. Michael's mama just gave me a funny look.

After dinner, Michael dropped his mama off then drove me back home.

"So listen," he said when we got inside. "There's something I want to show you."

"Okay."

"But I don't want you to get mad."

"Okay."

He led me to the kitchen, opened the drawer, and took out a knife and fork. "Now don't get mad," he said again, handing them out to me.

"Just say it, I won't get mad!"

"Okay, the thing is, you're holding the silverware all wrong."

"What you mean? How am I holding it?"

"You got it in a fist. You were gripping that fork at dinner like you were about to dig a hole and plant some flowers."

"No, I wasn't."

"Baby. *Trust me.*"

I looked down at the fork clutched in my hand. I felt my face burning hot with embarrassment. I wanted to punch Michael in the face for insulting me. I wanted to stick him in the chest with the fork in my fist. But instead I took a breath. "Okay," I said. "So how am I supposed to hold it?" He took my hands in his and showed me the right way.

We moved in together not long after that.

CHAPTER 22

Four More

I got the call at three thirty on a Tuesday afternoon. Michael had worked the early shift at the mattress factory and was coming in the door as I picked up the kitchen phone. On the other end of the line, my sister Sweetie was yelling like her house was on fire. "Rabbit! Girl, you gotta come over! The lady from DFACS say she gonna take away my kids!"

The last time I'd seen my sister was almost a year before, not long after Michael and I started dating. He said he wanted to meet my family, so I'd taken him over to Sweetie's place because she was the only one who wasn't in jail. We'd driven out to Lynwood Park, past streets lined with luxury homes, crossed an intersection, and suddenly we were in the hood. Sweetie lived with her three daughters in a broke-down house that reeked of stale Newports and dirty diapers. When Michael and I stepped inside I could see him, out of the corner of my eye, gagging from the smell. The floor was littered with overflowing ashtrays and dirty dishes crawling with roaches. In the middle of the room, turned on its side, was a wooden chair missing three of its legs.

I'd heard a rumor that Sweetie was hooked on crack. When I saw her that day with Michael—the way she stood at the window scratching and twitching and searching the road with paranoid eyeballs—I knew the rumor was true. Now here she was on the other end of the phone, begging for my help.

Talking a mile a minute, Sweetie explained that her case-worker had come by and told her the Department of Family and Child Services was taking her kids and putting them in foster care unless Sweetie could find a family member to take them. "She says I have until six o'clock, or else she's gonna get the police." Sweetie started to cry. "But I ain't an unfit mama. I'm a *good* mama. Rabbit, you gotta come get my kids."

"DON'T YOU GO GET THOSE KIDS," Michael said when I hung up the phone. He'd been standing in the doorway, listening to my call.

I ignored him and started pulling on my sneakers, my mind racing as I tried to remember how old Sweetie's kids were. There was LaDontay, who was a few months older than Ashley, so that put her at eight. And then two younger girls, Destiny and Diamond, who were two and three. But Sweetie had also been pregnant when Michael and I had seen her last. I did the math in my head and figured the baby couldn't be more than six months old.

I ran into Nikia's bedroom; I knew I had a box of old baby blankets in there, somewhere. Michael was right on my heels. "We can't take care of four more children," he said, watching me throw sneakers, water guns, and headless G.I. Joe action figures out of Nikia's closet, searching for the blankets. "You know that."

I turned around to look at Michael. "They're family," I said. "Where else they gonna go? Anyway," I added, turning back to the mess in the closet, "you like kids."

Ever since we'd been living together Michael had been acting exactly like a daddy. One day I came home after picking up Ashley and Nikia from school and found Michael had cooked us all dinner: fried pork chops, collard greens, and macaroni and cheese. Then he sat at the table and helped the kids with their homework.

"Your daughter can read," he said later that night. "But your son needs some help." I already knew Ashley was smart as a whip. Nikia was a different story. He was repeating kindergarten because he was so far behind.

"That boy takes after his daddy," I told Michael. "Derrick's stupid ass can't read, either."

"Let me try," said Michael. "I bet your son just needs a little extra help." He came home the next day with a bag from Kmart filled with a stack of reading books. "Come here, little man," he called to Nikia. "I got something for you." The two of them sat at the dining room table, Dr. Seuss's *Hop on Pop* cracked open in front of them. I stood in the doorway and watched Michael point to the words on the page. "Sound it out," he said gently. "Just take your time."

AT THE BACK OF NIKIA'S CLOSET, I finally found what I was looking for, a clear plastic box filled with Nikia's old baby clothes. I grabbed a blue flannel blanket and headed to the door. Michael was right behind me. "We need to talk about this," he said, following me out of the apartment, down the front walk, and over to my car. "I didn't sign up for all this . . . I'm serious. We got to think this through."

I opened the car door, slid behind the wheel, and turned on the ignition. I didn't understand why Michael was getting so worked up. Everybody knows that when your crackhead sister says DFACS is about to snatch her babies, you don't waste time thinking it through, you just go get them.

"There's nothing to talk about," I said as I pulled away from the curb. "I gotta get those kids."

In my rearview mirror I could see Michael standing in the road with his hands on his head. For a second he looked exactly like Curtis the day Mama put her dentures in the road and drove over them with her car.

SWEETIE'S HOUSE LOOKED WORSE than I could have imagined, it was like a tornado had lifted up a trash heap and dropped it right in the middle of her living room. My sister was standing at the window, holding a tiny baby and peering out the dirty bedsheets she'd hung up as curtains. On a sagging sofa against the wall were Sweetie's older girls. LaDontay was dressed in a grimy T-shirt and boy's shorts, but the other two wore only sagging diapers. All of them were covered in crusty silver-dollar-size sores.

Sweetie's caseworker was already there, standing by the front door, a manila folder in her hand. "I think the children may have ringworm," she said when she caught me looking at their scabs.

"My kids ain't got no worms!" shouted Sweetie from across the room.

The caseworker ignored her and turned to me: "Would you mind if we stepped outside for a moment?"

On the front porch, she cleared her throat. "I've offered to help find your sister a drug treatment program," she said. "Unfortunately, she's been very resistant."

"Okay."

"At this point, we are seriously concerned about the welfare of the children . . ."

"Yeah."

"My understanding is that you've offered to take them."

"Uh-huh."

"All four of them?"

"Yes, ma'am. They can stay with me until Sweetie gets herself together."

The caseworker lifted her eyes from her folder, looked at me, and smiled. "Wonderful," she said. "That's just wonderful." We made a plan that I'd take the girls for the night and the caseworker would come by my place the next day to walk me through the paperwork for getting temporary guardianship. As she turned to go back inside she added, "It's usually so difficult to find someone willing to take four kids at once. The girls are very lucky."

In the living room Sweetie was spinning around like a wind-up toy, picking up filthy baby clothes, mismatched flip-flops and broken toys, and shoving everything into a garbage bag. "I got the kids' things together," she said. "Clothes and whatnot."

"I don't need all that," I told her. "I'll get them some new clothes."

"But I got everything right here." Sweetie held out the trash bag and I watched a cockroach make its way up the side. "Take it!"

"Okay," I sighed, setting the bag down by the front door.

Then she handed me the baby. "Her name's Jonelle," Sweetie said. The girl was tiny, with milky eyes and dried mucus crusted around her nose. She smelled like she hadn't been changed all day. Sweetie turned to her older daughters on the sofa. She knelt down in front of them and opened her arms wide like she wanted to give them all a hug. LaDontay pushed the younger ones toward their mama but hung back on the sofa, her eyes darting between me, the caseworker, and her mama, like she was trying to figure out which one of us she should trust. She bit her lip and her eyes filled with tears.

Sweetie didn't notice. She clutched Diamond and Destiny. "Y'all be good for your auntie," she said, her voice cracking. "I'm gonna see y'all soon. I just need to clean up the place a little. That's all. I just gotta clean up."

"Okay, c'mon, let's go." I held out my hand for the girls. "Tell your mama you love her. We gonna go see your cousins now."

I left the trash bag full of filthy clothes and took the dirty kids. Sweetie stood in the middle of her living room, surrounded by garbage, and watched us go.

WHEN I WALKED IN MY FRONT DOOR, I was relieved to see my niece Cece had come by and was sitting on the sofa, watching TV. "Thank you, Jesus," I said under my breath. Then I called to her, "Girl, get over here. I need some help."

Cece stood up, holding her nose. "What's that *smell*?"

I didn't answer, I just handed her Jonelle, whose shit-soaked diaper was hanging off her like a second booty. "These are Sweetie's kids," I explained. "They're gonna stay with us a while."

"What the hell am I supposed to do with *this*?" Cece asked, holding Jonelle away from her body like I'd handed her a ticking bomb.

"Use your head!" I answered. "These babies need a bath." I told Cece to put all the girls' clothes in a trash bag and tie up the top. "And make sure they don't touch nothing," I added. Then I went to look for Michael to ask him to run to the store for diapers and formula.

I found him in our bedroom, sitting on the edge of the bed, pulling on his shoes. On the floor in front of him was his khaki green army-issue duffel bag, packed up and cinched at the top, like he was moving out.

I knew Michael was mad when I left for Sweetie's, but I figured he was just being dramatic, like Denzel, putting on a show to make his point. "What the hell?" I said eyeing his packed bag on the floor. Michael didn't say a word. He just stood up, grabbed his duffel bag, and slung it over his shoulder.

"Wait . . ." I said. "Are you *leaving*?"

"What does it look like?" he said, walking past me and heading for the door.

"Wait!" I called after him. "You asking me to choose between you and those babies? Is that what you want?"

"I'm not asking you to choose," he said without turning around. "I'm just telling you this is too much."

I followed Michael down the front walk and out to his Nissan. He opened the door and got inside. Without thinking, I ran in front of his car with my hands in the air.

"Don't go!" I yelled.

"Move out the way, Pat."

"You can't leave!"

Michael started his car.

"Don't make me choose!" I yelled again. I could see him behind the wheel, stony faced and staring straight ahead. I realized at that moment I would do *anything* to make him stay.

Michael and I had been living together almost a year and I'd come to count on him being there like the sun in the sky. No matter what happened on Ashby Grove—if it was a slow night, or I just didn't feel like going, or even the time I dozed off at my friend Tanisha's house and got robbed by her crackhead brother who stuck his hand down my bosom and stole fifteen hundred dollars in cash—Michael held us down. He made sure the bills were paid on time and there was always food in the fridge. It was his idea for us to eat dinner together every night and read stories to the kids before they fell asleep. Michael was solid and stable, and with him everything was better. There was no way I was gonna let him go.

I stepped one foot onto the front bumper, then heaved myself onto the hood of his car. "Please don't go," I cried, banging my fists on his front window. *"PLEEEEEASE!"*

Michael was ex-military, but I guess even combat training didn't prepare him for the sight of me spread eagle on top of his car. He turned off his ignition and stepped onto the curb. "Golly," he said. "You keep making all that racket, the neighbors gonna call the police. You need to get off that car and get back inside before you get arrested for causing a disturbance."

"I'm not going unless you come with me," I said, not moving. Michael grabbed onto the belt loops of my Levi's and started sliding me off the hood of his car.

"Nooo!" I wailed, clinging to his wipers. "Say you'll come inside."

"All right," he said, finally. "But I'm only staying *one day*. I'm leaving tomorrow, Pat. I mean it. One day."

IT TOOK ME AND CECE HOURS to get the children clean. We scrubbed their bodies, cleaned their ears, cut their nails, and washed all their hair. While we were busy in the bathroom, Michael went to the corner store for diapers. When he came home, he made a big pot of spaghetti. We fed the kids dinner, put the older three girls to sleep on blankets on the floor in Ashley's room, and took Jonelle into bed with us, wrapped in one of Michael's T-shirts.

In the dark, Michael said again we couldn't keep the kids. He said the same thing the next day when he came home from work. He said he didn't want all the responsibility, and there was no way they could stay.

But later that night, I put Jonelle in his arms and he rocked her to sleep. He read the girls a bedtime story. He helped me rub ointment on their ringworm and grease their scalp.

A week after they arrived, Michael came home from work with a shopping bag from Kmart. Inside were four *Little House on the Prairie*-style dresses, covered in ruffles and bows. I didn't tell him they were the most ugly-ass dresses I'd ever seen. Instead I just threw my arms around his neck and kissed his face. "I knew you'd come around," I whispered.

"Seriously, Pat, we really need to talk about this."

"They're not going to be here long," I promised.

Ten years later, the girls were still with us.

CHAPTER 23

Letting Go

Not long after Sweetie's girls came to live with us, her caseworker gave us a Section Eight voucher to help with the rent. That changed everything. It meant we could move all the kids into a bigger place. Michael and I spent almost two weeks driving through Atlanta, Decatur, and Marietta looking for a house to fit six kids. One place out by the federal prison was so run down—with holes in the drywall and electrical tape holding together broken windowpanes—that Michael took one step inside and turned right back around. Another place was clean enough, but I noticed a couple of dope boys slinging rock on the corner and told Michael, "We ain't bringing the kids into this mess."

Every place we saw was worse than the last. Then we found it: a clean, four-bedroom house on a quiet block in Riverdale, a middle-class suburb about half an hour outside Atlanta. We signed the lease and moved right in. On our first night, Michael and I lay in bed listening to the crickets outside our window. "This is going to be nice for us," he said. "It's like we're getting a brand-new start."

"Yeah," I agreed. "It's good out here."

Michael was quiet for a while. But just when I thought he'd fallen asleep, he said, "Now that we out here, maybe it's time you started thinking about getting yourself a real job."

"I got a job," I reminded him. "I'm an entrepreneur."

"No. I mean something *legal*." That's how I found out that when Michael said we were "getting a brand-new start" he meant a new start for *me*.

I thought we had an understanding. I understood that he went to work every day at the mattress factory and paid the rent. And he understood that I sold crack and bought us nice shit. I knew Michael wasn't crazy about me being a drug dealer, so when I'd come home with a brand-new big-screen TV it was pretty much a don't-ask-don't-tell-type situation. He didn't ask where the set came from and I didn't tell him about my hustling. That was our arrangement. But now Michael wanted me to get off the streets and straighten up my life.

He started bugging me about it all the time. "Is this how you want to live?" he'd ask. Or "Just put in an application anywhere, get your foot in the door." Sometimes he'd even try to scare me: "What if you get busted? None of these kids have my last name. If you get locked up, DFACS is gonna take all six of them. Is that what you want?" Michael wouldn't let up, and I wouldn't back down. Every time he complained about how I was making money, I'd storm out of the house and go shopping. Pretty soon we had all brand-new furniture.

As much as I hated Michael telling me what to do, I had to admit he had a point. Selling dope wasn't a good fit for our new lifestyle. One minute I'd be taking Nikia to Little League practice, dressed in a button-down blouse and leather flats; the next I'd be rocking jeans and Jordans and running out to the hood. I was living a double life. But it was more than that. Living in Riverdale forced me to finally face the truth. I couldn't keep telling myself that selling dope wasn't hurting anybody.

All I had to do was look around. Our new neighborhood wasn't filled with crackheads, hookers, and dealers. Riverdale was home to everyday folks who got up, went to work, came home, and played with their kids. It got me thinking about the time, before crack came to town, when Ashby Grove and Baldwin Place were filled with these same types of people. When I first started hustling, I used to tell myself that crack was just another high. But after six years serving addicts, it was obvious; crack ruined everything it touched.

Sweetie had her babies taken away behind that shit, and she wasn't the only one whose life was ruined. Sometimes I'd lie in bed and think about all the folks I saw lose everything after they got hooked.

I thought about this one lady who used to come by Ashby Grove looking for her daughter. She showed me a picture of a round-faced girl about my age. "She was a history major at Spelman," the mama said. "Have you seen her?" I glanced at the photo and told the lady, "No, ma'am." But later it hit me: the girl in the picture was Butterfly.

ONLY SOMEBODY WHO'S NEVER HUSTLED BEFORE would think you can go from slinging crack to becoming a law-abiding citizen overnight. But that's not how turning your life around works. It's a *process*. It takes time. Especially if you're going from making easy money to minimum wage. Michael wanted me to give up hustling and get a regular job, but quitting cold turkey would be a shock to my system. I had a better idea. I figured I should ease out of selling drugs and into something less risky. Lying on the sofa one afternoon watching *The Young and the Restless* while the kids were at school, it came to me in a flash: I could run a scheme like Brenda, my idol from Fulton County Jail. Instead of selling drugs in the hood, I could write phony checks at the mall. Except for Brenda, I'd never heard of anybody getting busted for check fraud. It was the perfect scheme, upscale *and* low risk.

I gave the rest of my dope to my niece Cece so she could make herself some spending money, and settled up my debt with my dealer Lamont. "I'm proud of you, girl," he said when I told him I was quitting the game. "I always knew you were better than this."

I was turning my life around, giving up slinging rock to start my classy new enterprise, writing bad checks. I thought for sure Michael would be happy about me taking my hustle in a new direction. But if anything, he was madder than before. "You can't keep doing this," he said one evening when I walked through the front door after a hard day of work scheming at the mall. I was carrying five overflowing shopping bags from Macy's, Foot Locker, and Gymboree, all paid for with forged checks.

"But it's school clothes for the kids!" I insisted.

"That doesn't make it right," he said. Then he walked into the bedroom and slammed the door behind him.

I couldn't believe Michael was giving me a hard time when he should have been *thanking* me. Running bad checks took a lot of planning and coordination. Every week I needed to get my hands on a fresh set of stolen checks, which I bought from a girl I knew who worked at the post office. She would intercept boxes of checkbooks that the bank had mailed to its customers then sell them to me for two-hundred-dollar Macy gift cards. I also needed fake picture IDs to match the name on the checks. I got the ID from a brother who made them in the back of his poster store in a janky-ass strip mall out in Decatur. Then I had to look the part, dressing in dark slacks, a conservative blouse, and leather flats, strolling the mall with confidence like I really did have a bank account full of money.

Every time I made a purchase and paid for it with a fake check, I felt invincible, like I could do anything. I especially loved spending my fake money at stores where the salespeople used to follow me around and give me the side eye when I came in dressed in Jordans and jeans. It was a real eye-opener. I didn't realize how prejudiced the salesladies at Macy's were against drug-dealer money until I started doing white-collar crime.

I put a lot of work into my operation, but Michael didn't seem to appreciate any of it. "It's not right," he kept telling me. "You're stealing from folks."

I tried to win him over with footwear. I bought him Timberland boots, some Grant Hill high tops, a pair of black Charles Barkley Air Max with a red Nike swoosh, and a luxurious pair of blue snakeskin cowboy boots. I came home from the mall and laid them all out on the coffee table in front of him. All he said was "What I'm gonna do with all this? I only got two feet."

"What I don't understand is why you aren't even *trying* to get real job," he said a few nights later when we were lying in bed having another version of the same conversation we'd been having for months. "You're a people person. You could do all kinds of things. Like, I could see you working at a car dealership. With your personality, you could sell the heck out of a car. Isn't there anything you want to do besides collecting a welfare check and running schemes?"

I was quiet for a long time. "I don't know what else I *can* do," I said, finally. "What if hustling is the only thing I'm good at?" But Michael was already sound asleep.

It felt like hours that I lay in bed staring at the ceiling. Maybe Michael was right. Maybe I *wasn't* trying to do better. Michael wanted me to turn my life around—not just for myself, but for him and the kids, too—and I was busy chasing the rush I got every time I flashed some fake ID and walked out of a store with hundreds of dollars' worth of merchandise. I was getting high off the thrill. As much as I blamed Sweetie for choosing drugs over her babies, I wasn't any better. The only difference was that when they were with her the kids had ringworm, with me they were covered in stolen clothes.

I didn't tell Michael I decided to stop hustling. I just started going to the mall less and less often. Then one day, I threw my stolen checks away.

Later that evening, Michael found them in the kitchen trash. He pulled out the checkbook and brushed off a strand of spaghetti. "What's this?" he asked.

"Those are mine. I'm not doing that shit no more."

He peered at the check in his hand, then up at me. "*This* is who you've been pretending to be?"

"Yeah."

He let out a long whistle. "You are damn lucky you didn't get caught."

"What you mean?"

"Pat, you're a lot of things. But I can tell you one thing for sure, you *definitely* don't look like no 'Mrs. Bella Bernstein.'"

CHAPTER 24

Job Readiness

S tocking shelves at Target during the overnight shift was the first job I got fired from. But it wasn't my fault. My shift started at midnight and during my lunch break I went to sit in my car. Of course I fell asleep. It could have happened to anybody, twice.

After that I got a job as a cashier at a Texaco station. The manager fired me when he caught me on video surveillance using my kids to help stock the cooler with soft drinks. That wasn't my fault either. How was I supposed to know there's a company rule that says only actual "employees" are allowed to do the work? Then I got a gig as a cashier at Walmart. My supervisor asked me to stay late and cover for another employee who'd called in sick. I thought she had a lot of nerve, so I told her to kiss my ass. She let me go, too.

That's when it really began to sink in that there aren't a lot of job opportunities available to a twenty-three-year-old never-went-to-high-school former drug dealer. I was running out of options, so one afternoon I drove to the next town over, where nobody I knew would see me, and applied for a job at what had to be the slowest McDonald's in all of Georgia. The manager hired me on the spot.

Except for the before-work morning rush and the high school kids who came by after school, we hardly ever had any customers. The only thing that kept me from dying of boredom was Cindi, the little white girl who worked the counter beside me. Cindi had more problems with her love life then the entire cast of *The Young and the Restless* put together.

One afternoon a few weeks after I got hired, Cindi was deep into one of her stories—"So I told Travis, 'Go ahead and take your skanky ex to prom.' He's such a dumb-ass he don't even know they don't make *maternity* prom dresses . . ."—when I glanced up and noticed a white man dressed in a crisp dark suit had stepped inside. He looked around the dining area, peered into the restroom, then leaned forward and talked into a little microphone pinned to his lapel: "All clear!" Two more white men walked in behind him. One hung back by the door, the other stepped to my register. He looked familiar.

"What can I get for you?" I asked.

"I'll have a cheeseburger, a cup of water, and a side salad."

He was an older dude with gray hair and a friendly smile. He looked a little like Bob Barker from *The Price Is Right*. I was *sure* I'd seen him on TV. I thought maybe he was on one of those white shows Michael was always watching, like *Seinfeld* or *60 Minutes*.

I couldn't place his face, but I was pretty sure the man waiting for his burger was some kind of famous. The curiosity was killing me, so I pointed my finger at his chest: "Nigga, where do I know you from?"

Beside me I heard Cindi gasp. "Pat!" she whispered, loud enough for everybody to hear. "Girl, that ain't no *nigga*. That's Jimmy Carter. He used to be the *president*!"

I turned to Cindi, whose eyes were bugging out of her head, then back to Jimmy Carter, who looked almost as shocked. I couldn't help myself, I bust out laughing. "I *knew* I recognized your ass!" I said. "I'ma give you your cheeseburger for free!"

A few weeks later, after my manager told me to wipe down

some tables that were already clean, and I told her she must be out of her damn mind, I got fired from McDonald's, too.

In less than a year, I'd had five jobs and lost all of them. It was getting hard to stay positive. Sometimes, I got to thinking that maybe it was *me*. Maybe I just wasn't built for legal employment. Michael tried to make me feel better. Whenever I got fired—or, as I explained it to him, "I quit"—he always gave me the same pep talk: "Nobody in your family ever had a regular job. You just gotta get used to the lifestyle!" Even so, I was beginning to wonder if I'd ever be able to hold it down.

IT SEEMS LIKE EVERY TIME I'M AT MY LOWEST, somebody comes along to pull me back up. Miss Troup, Brenda and Eva, Hubert Hood, all of them were like my personal cheerleading squad telling me to keep up the fight when I thought I was losing the game. In 1997 Miss Campbell came into my life like a star cheerleader doing back flips down the field. She was a caseworker for the city of Riverdale's Positive Employment and Community Help. The program, called PEACH for short, was part of President Bill Clinton's Welfare-to-Work plan, which was supposed to get folks off welfare. The way Miss Campbell described it made it sound like a prize.

"What are you interested in, Patricia?" Miss Campbell asked at our first meeting. She had a warm smile and was full of enthusiasm. Whatever I wanted to do, she said, there was job training to get me there. She riffled through a stack of papers on her desk: "Are you interested in training to become an office administrator? Or perhaps a cosmetologist? Or how about this . . ." She pulled out a folder and flipped it open. "A medical assistant? It says here you'd be qualified to work in a doctor's office, health clinic, or hospital."

How did Miss Campbell know I'd always dreamed of having the kind of job where you got to wear scrubs to work every day?

Even after I quit forging checks, I still kept my fake "Grady Hospital Staff" ID in my wallet. I really liked the idea of helping people.

"Yes, ma'am," I said. "Medical assistant, I want to do that."

IT TURNS OUT THAT GETTING a medical assistant certificate is *not* easy. First I had to get my GED, which meant learning all the stuff—eighth grade all the way through high school—I'd missed while I was slinging dope and raising kids. That took six and a half months. Then I had to enroll in a nine-month medical assistant program at a night school located in an office in the back of a strip mall. When I went to sign up, the lady at the front desk handed me a stack of papers to fill out so I could get the seven-thousand-dollar student loan I'd need to pay for the class. Nowhere on the forms did it say it was going to take me almost twenty years to pay off the loan.

I went to Medical Assistant School five days a week, and learned how to draw blood, weigh babies, give vaccinations, and take blood pressure. On February 11, 1997, the school held a little graduation ceremony. Michael and our six kids all cheered when the program director called my name. Even Miss Campbell came to the ceremony. "I'm so proud of you," she said, giving me a hug. "You've worked so hard!"

After Medical Assistant School all I had left was a four-week class at the PEACH office, called Job Readiness. I joined a group of other unemployed "ladies and gentlemen," as the instructor liked to call us, to learn important skills, like "dress for success," which, lucky for me, turned out to be the exact same outfit as the one I wore passing bad checks at the mall. It took the girl with the giant pawprint tattoos across her bosom three separate tries to get the look of "appropriate office attire" right. One day she showed up in four-inch heels like she was hittin' the club; the next day she came in a yellow satin church hat, which I'm guessing she got by jumping an old lady on Sunday morning.

I'd taken three Job Readiness classes when I realized I was *already* ready. I started searching help-wanted ads in the back of the *Atlanta Journal-Constitution* and found what I was looking for in the Monday paper, right at the bottom of the page. A doctor's office out in Sugar Hill was looking for an assistant. I knew the area; I'd driven through it with Lamont. It's what he called "quality."

"This is it," I said to Michael, showing him the ad. "This is the perfect job for me."

I PULLED UP TO THE DOCTOR'S OFFICE twenty minutes early for my interview, my stomach doing flips. At five to nine, I slipped out of the car, smoothed the front of my black slacks, and walked inside.

The office manager, Miss Shelly, sat behind her desk and asked me all the questions I'd already practiced in Job Readiness: What's your greatest strength? *I'm a quick learner!* Where do you see yourself in five years? *Working here, I hope!* What's your greatest weakness? *Sometimes I just care too much!*

She asked if I had any children. When I told her I was raising six kids, her face lit up. She had four of her own, she said, and proceeded to tell me all about them while I nodded and smiled. On her desk was a picture of her four-year-old daughter, Christy, dressed in cowboy boots and a Dolly Parton wig, looking like she was next in line for JonBenet Ramsey's killer.

"My baby girl got second place prize in the Miss Precious Baby Peach Pageant," said Miss Shelly, with pride, when she saw me glance at the photo. To me, her baby looked like a miniature hooker. But I said, "She's real pretty, ma'am. She should have won first place!"

I guess complimenting Christy was part of the interview, because Miss Shelly offered me the job. "Hon," she said, "you have such a sunny disposition, I'd love to bring you on board!" I was going be making twelve dollars an hour, plus benefits; Michael was going to be so proud.

Miss Shelly stood up and walked around her desk to give me a hug. "There's just one more thing," she said. Then she dropped the bomb.

She wanted me to go to the police station down the road and ask the officer at the front desk to run a criminal background check. "It's nothing," she said with a wave of her hand. "Just a formality. After you bring back the paperwork, we'll sit down and look at the schedule. I want to get you started right away."

IT HAD BEEN SEVEN YEARS since Officer Harris had busted me on Ashby Grove. I'd done my jail time, cleaned up my act, and gone back to school. As far as I was concerned, I'd put the criminal part of my life behind me. It was history I wanted to bury. But I guess there are some things the world won't ever let you forget.

When I handed my ID to the officer, I mentally prepared myself for what I was about to receive: a piece of paper detailing my charge of "Possession with Intent to Distribute." I didn't want to see it, but I knew I had to face the facts.

I almost passed out from shock when instead of a single sheet of paper, the officer handed me a report almost twenty pages long. "Hang on," he said, passing me the stack. "I gotta go put some more paper in the machine."

I took the report out to my car and read it over, my heart sinking lower with each turn of the page. There were charges from incidents I barely remembered, like "Abandonment of Certain Dangerous Drugs" in 1991, and, three years later, in 1994, a charge for "Financial Transaction Card Theft" for the time I went on a shopping spree with a stolen credit card DeMarcus had traded me for a ten-dollar rock. The most recent charge was only a year old: a 1996 misdemeanor for "Depositing Bad Checks." The charge sounded a lot worse that it was. I imagined myself telling Miss Shelly: *I can explain that one! Michael's paycheck wasn't gonna come till Friday. But it was Wednesday and the kids were hungry, so I wrote a check at*

the grocery store for twenty-seven dollars and thirteen cents on my
empty bank account. But I paid it all back!

But who was I kidding? Even if I could explain why I wrote
one bad check, it wouldn't make a difference. I had twenty pages
saying I wasn't to be trusted. And in case Miss Shelly wasn't sure,
right there on the first page of my Criminal History Report were
the worst words of all: "Convicted Felon."

I felt like I'd been blindsided. Nobody at medical assistant
school asked me if I was a convicted felon before they took my
money. None of the instructors at Job Readiness ever brought
this up. Maybe if I'd heard, even one time, "criminal background
check," I would have known this was coming instead of getting my
hopes up, like a fool.

I sat in the car for almost an hour staring at the pages until
they began to blur. All that work, all that studying, all that time
and effort and student loan money. Why did I even bother? Miss
Shelly would never hire somebody like me. Nobody would. I felt
stupid for even trying.

I went home and told Michael the position had already been
filled.

WHEN I TOLD MISS CAMPBELL WHAT HAPPENED, sitting in her
office later that week, she took my hands in hers. "I'm so sorry,"
she said, leaning forward. "That must have been very difficult."

"I was *so* embarrassed," I said. "I almost threw my jacket over
my head and ran out of that police station. But I just got my hair
fixed, so you know I wasn't trying to mess up my weave."

She let out a little chuckle. "Of course."

"It's not like I don't know I've done wrong in my life," I continued. "But I swear, I never even *seen* some of those charges before."

"Really?" Miss Campbell sat up in her chair, looking concerned.
"If there's been some kind of computer error on your records, we
should look into it and have it corrected."

"Yeah!" I said, getting excited by the thought. If we could erase some of these charges, I'd have a chance! I pulled the papers out of my purse and scanned the first page. "Well, like this one here for assault . . ."

"Yes?" Miss Campbell leaned forward. I could tell she wanted my criminal background to be a mistake as much as I did.

I carefully reread the charge. "Actually," I said, looking up from the page and biting my lip, "come to think of it, I *do* remember this. It's from the time I hit one of my customers with my car. . . Yeah, okay, maybe I *did* do that." I turned the page. "Okay, so this one right here! It says 'Insulting or Abusing Public School Teachers, Administrators, or School Bus Drivers' . . ." I paused, remembering an incident that had happened shortly after Michael and I moved to Riverdale. I'd gotten onto a school bus and cussed out the driver for driving off the day before without my kids because they weren't at the stop on time. All I did was very carefully explain to the driver, "Do that again, bitch, and I will knock your gotdamn block off."

I looked up at Miss Campbell. "Okay, never mind that one. I just didn't know it was gonna be on my permanent record . . ."

I turned the page again. "Okay, *this* I didn't do!" I tapped the paper with my finger. ". . . Oh wait. Maybe I did do that."

I glanced at Miss Campbell. She looked stunned. I cleared my throat and started neatly folding the papers and putting them back in my purse. "You know, now that I'm giving this a closer look . . ." I shook my head and laughed. "I guess I wouldn't hire me either. I mean, *damn*. I look like a career criminal!"

Miss Campbell's mouth was beginning to curl up in a smile. She shook her head quickly, like she was trying to shake herself serious. But it didn't work; she started to laugh, too.

Ever since I first met Miss Campbell I noticed she wasn't like any other caseworker I ever had. Those other ladies would listen to stories of my childhood, clutch their hearts, and tell me how sorry they were for all my troubles. When I told Miss Campbell how Mama taught me to pickpocket before I learned to read, or how Derrick had more baby mamas than most men had teeth, she cracked up. Sometimes she got to laughing so loud she had to close her office door because the other caseworkers were complaining about the noise. The more she laughed, the more I wanted to make her laugh.

"Twenty pages!" I howled. "It was thicker than a phone book. When that officer handed me my Criminal History, I thought he was gonna arrest me on the spot for using up all his ink." The next thing I knew, the two of us were cracking up so hard we could barely catch our breath.

"Oh, Patricia," Miss Campbell said, wiping tears from her eyes. "Honey, you're out there trying to be a medical assistant, but I think you missed your calling. The way you turn a sad story around, you should be a comedian! You're the funniest person I know."

CHAPTER 25

Eight Minus Four

Sweetie hardly ever came to see her girls. By the time they'd been with us for almost ten years, she'd only visited a handful of times. She'd show up, swear she was off drugs, and then months, even years would pass, before she'd come back. Every time she disappeared, I'd have to deal with the girls' broken hearts. "Don't worry about it," I'd say, patting the back of whichever one of them was most upset. "You'll see your mama when you see her." But the girls weren't stupid. They knew Sweetie was choosing drugs over taking care of them.

I tried to teach my nieces that there was no point thinking about shit you can't change. That's how I survived. If I had a bad feeling, I pushed it away and kept it moving. That was the main difference between me and my sister. Sweetie and I had the same shitty time coming up—we both went hungry, got beat, and had Mr. John touch on us—but I shoved down those memories while she chased them away getting high.

I thought if I could teach the kids to ignore their sadness, they'd be okay. But Jonelle was so little. Whenever Sweetie would show up only to vanish again, she'd cry her eyes out and start wetting the bed. Destiny would stop talking and Diamond would start acting up at school. Sometimes I wished Sweetie wouldn't come around at all. Without her interrupting, our lives were pretty good.

Michael and I were raising six kids. Most people couldn't handle all the noise and commotion, but I loved it. The laughing and arguing, the clothes all over the floor, the missing toothbrushes, the shoe with no match, and the never-ending list of things I had to do to make sure everybody was happy, clean, and fed, being a mama to all those babies was everything I'd ever dreamed. That's why when Jonelle, Sweetie's youngest, was two years old, Michael and I decided to have one more.

FROM THE SECOND I GOT PREGNANT, I'd never seen Michael so happy. "She's got my nose!" he exclaimed, grinning and pointing at the screen when we went for our first ultrasound.

I looked over at the technician and rolled my eyes.

"Sir," she said, "that's the baby's foot."

Our daughter was born March 25, 1998. We named her Michaela, for her daddy. After we brought her home, all Michael wanted to do was hold her in his arms and sing K-Ci & JoJo's "All My Life" over and over.

"You're making me jealous," I said one night when he wouldn't come to bed. He just laughed and kept on singing. "I mean it!" I yelled, stomping out of the room. With all the love and affection he was pouring on that little girl, it's a miracle I even got Michael to look my way. But I guess being a daddy put him in a baby-making mood, because eight months later, I popped up pregnant again. This time we had a boy. We named our son after his daddy, too, Michael Jr., but everybody called him Junebug.

Then we had eight; Junebug in diapers, LaDontay and Ashley in high school, and the rest of the kids filling the space in between. We put the older kids in every kind of after-school activity—track, cheerleading, football, and baseball. Ashley played viola in her high school orchestra and LaDontay was in the ROTC. I went to PTA meetings, volunteered for school field trips, and spent hours standing on the sidelines, in the sun and the rain, cheering for whichever kid was on the field.

At night I put all the kids to bed, just like June Cleaver, turning out the lights: "Good night, Ashley. Good night, LaDontay. Good night, Diamond and Destiny. 'Night, Jonelle and Michaela. Sleep tight, Nikia and Junebug. I love you!" Then I'd climb into bed beside Michael and listen to him snore like a gotdamn freight train pulling into the station.

LADONTAY RAN TRACK. She was good, too. The summer after she finished eleventh grade, she was invited to a meet in Florida to qualify for the Junior Olympics. Michael and I weren't about to let her go down there by herself, so we packed everybody into the back of our Montero Jeep, drove six hours to Orlando, and checked into the Marriott hotel. The trick to having eight kids in a single hotel room is you can't walk in with them all at the same time. You have to send them through the lobby and past the front desk one by one, telling them to run when the clerk looks the other away.

Back then, Michael and I were both working the assembly line at General Motors. Before the trip, we did two months of overtime, scrimping and saving almost five thousand dollars between us so we could show the kids a good time. In Orlando, we took them to the beach, the movies, and the all-you-can-eat buffet. When we got home, Michael pulled out his wallet. Thirty-four dollars was all we had left. It was worth every penny. That trip was the last time we'd ever have that much fun with Sweetie's four girls.

THE OLDER KIDS WERE AT SCHOOL and I was home with the babies one morning, when the phone rang. It was Sweetie on the other end of the line. I felt my heart stop when she told me why she'd called. "Rabbit," she said, "I'm coming to get my kids."

Sweetie had gone to family court, she explained, and filled out all the paperwork to take away my temporary custody. I guess she knew by the way I hung up on her that I'd put up a fight, because when she turned up at my house later that day, she brought along the cops.

We all crowded into the kitchen: me, Michael, Sweetie, two police officers, and Miss Campbell, who rushed right over after I called her in tears. The kids were upstairs and Michael had his hand on my shoulder, trying to keep me calm. "Why are you doing this?" I asked Sweetie. "You know you can't take care of them."

"They need to be with me. *I'm* their mama," she said, folding her arms in front of her chest. "Besides, I got the papers. You can't stop me." I stared at my sister, trying to think of something to say to make her change her mind. Maybe if she knew how much work it took. It wasn't just the times the girls were sick, or crying, or bickering with each other, stomping on my last nerve. Parenting required shit Sweetie would *never* even think of. Like the time I had to run to school on picture day and beg the photographer to do a group shot with four kids at once because I couldn't pay for eight different photo packages. Or the time I had to get LaDontay a date for the ninth-grade dance. It wasn't just any dance, it was the ROTC Military Ball. Ashley was going with our neighbor's son, but LaDontay didn't have a date. *I* was the one who saw the nice-looking boy working behind the counter at Subway and said, "You want to take my niece to the dance?" *I* hired a limo. *I* bought LaDontay and Ashley sequined dresses and had them looking like superstars when the boys came to pick them up. Sweetie would never do all that. She had her chance to be a mama to her girls. She was a fool for missing out.

"How you gonna take care of them?" I asked. "You still getting high . . ."

"Nah, Rabbit," she said, cutting me off. "I don't do that no more. I cleaned my shit up. I got my own place, and I want my kids."

Looking at her standing in my kitchen, waving the custody papers in my face, all I could think of was the night I picked up her babies to come live with me. In my mind's eye, I could still see Sweetie kneeling in the middle of her filthy living room, trying to get her daughters to give her a hug. I remembered the way the girls were covered in sores, with no shoes on their feet. I pictured little Jonelle, so tiny and helpless. In that moment I realized it didn't matter if Sweetie swore on a stack of Bibles that she would spend the rest of her life taking care of those girls. It didn't matter because when I looked at my sister, all I saw was Mama. That's when I lost it.

I lunged at Sweetie, my arms swinging like windmills, trying to slap the shit out of her. The policemen reached out to pull me off, but Michael grabbed me first. "It's okay," he said, pressing my face to his chest. "It's okay."

He leaned down, held his mouth close to my ear, and whispered to me, "They're not your kids. You've got to let them go."

In the weeks after the girls went to live with their mother, they called a few times to tell me they were hungry and there was no food in the house. I brought them groceries and took them to get their hair done. But after a while, the phone stopped ringing.

Maybe I should have fought harder to get them back. But it felt like nobody was on my side. Dre, Sweetie, Miss Campbell, even Michael kept telling me the girls needed to be with their mama. I'd done everything I could to raise my nieces the right way, but once their real mama popped up, it was like I didn't even matter. *Fine,* I thought. And I gave up the fight.

THAT JUNE, ASHLEY GRADUATED FROM Riverdale High School. Michael and I sat in the bleachers on the school's football field, sweating through our good clothes in the noonday heat. This was the moment I'd been dreaming of since the day Ashley was born. No one in my family had graduated high school. Not me, not Mama, not Sweetie, not Aunt Vanessa or Uncle Skeet, or Dre or Andre or Jeffro or Granddaddy. My baby girl was the very first one.

I did it, I whispered to myself as Ashley walked across the stage in her cap and gown. I looked around at the other families seated beside me in the stands. There were daddies cheering their children's names, and mamas smiling and dabbing their eyes with neatly folded tissues. I was the only one sobbing uncontrollably, tears and mascara running down my face.

In the car driving home later that afternoon I turned to Michael. "I keep thinking about Sweetie's girls," I said. I had big plans for my nieces. They were all going to graduate high school and make something of themselves. When Sweetie took her daughters back, I'd bawled my eyes out for weeks. It was like someone had snatched my own children away. "I don't know what's going to happen to them now."

LaDontay had been halfway through her final year of high school when she went back to her mother's house. She was the only one of Sweetie's girls to graduate. The next year, she was pregnant. Soon after, Destiny had a baby, too. When Jonelle, Sweetie's youngest, turned up pregnant at thirteen, Sweetie sent baby-shower invitations to Jonelle's entire seventh-grade class.

CHAPTER 26

Angels

It's funny how you can spend your whole life chasing a dream only to find out when you get it that it's just not enough. When I was living in Granddaddy's liquor house, I'd fantasize about the kind of life I saw on TV. I wanted checkerboard curtains in the kitchen and a man who came home every night. I dreamed of a clean home, a fridge full of food, and children who hugged and kissed me and told me I was loved. Living in Riverdale with Michael and our kids, I finally had everything I'd hoped for.

Michael and I cheered on Nikia at football and helped Junebug learn to read. We got Ashley off to college and tried to get Michaela to play baseball, even though all she just wanted to do was dig in the dirt and complain about it being hot. Even after Sweetie took her girls, life was better than I'd ever imagined. But still, something wasn't right.

It's hard to put into words the feelings I had back then. All I can say is it reminded me of how I felt those days when I missed Free Breakfast at school and had to wait for hours for the lunch bell to ring. I had a hunger that gnawed at me like an empty belly. I knew I *should* be happy. For the first time in sixteen years I wasn't taking care of other people's kids, ripping and running, hustling to get by. But at night I'd lie in bed feeling the hunger taking over. Even Michael noticed something was wrong.

"What's going on?" he asked, standing in the kitchen one evening, watching me wash the same dish over and over like I was in a daze.

"I don't know," I told him. I wasn't sure what I needed to fill me up. I didn't even know what I was craving. It wasn't until months after Sweetie's girls moved out that I finally found what I was missing.

I don't know what the hell told me put my name on the sign-up sheet for the open-mic comedy show at a little neighborhood bar that night. My caseworker, Miss Campbell, had told me I should be a comedian all those years ago. Maybe part of me was curious to find out if she was right. Or maybe I just wanted the attention. All I know is once my nieces were gone, and the house got real quiet, I found something pulling me to the stage.

I only had one joke. Technically, it wasn't even a joke. It was just me talking about my brother Dre breaking into folks' houses. "So, my brother is a cat burglar," I began, gripping the mic. "So he's a cat burglar, but he's fat as hell. He's a muthafuckin' fat cat burglar." There was no setup, no punch line. It was just me talking about Dre until my five minutes were up. Standing at the front of the room, with the noise and laughter and all those folks listening to my story and smiling back at me, I finally felt full.

I REMEMBER MY MAMA WAS ON TV ONCE. It was back in 1980, a few months before Curtis walked out on us. We were still living in the shotgun house on Oliver Street. I was eight years old and should have been at school, but instead I was home that day with Mama, the two of us glued to the set watching *The Young and the Restless,* when we heard a knock at the door.

Mama told me to go answer. When I flung open the front door, I was shocked to find Miss Monica Kaufman from *Action News* standing on the porch. Miss Kaufman was famous. She was the only black woman I'd ever seen on TV reading the nightly news and the only reason Mama tuned in.

"Hello," Miss Kaufman said, leaning down to talk to me. I marveled at her perfect teeth and her baby-blue pantsuit. "What's your name?"

"Rabbit."

"Well, hello, Rabbit! It's a pleasure to meet you, sweetie. Do you think you can go get your mother for me?"

"Yes, ma'am!" I ran to get Mama, then followed her back to the door so I could listen in on the conversation. Mama was dressed in a housecoat and head scarf, and kept reaching her hand up to brush her face, as if to smooth away a stray hair. I guess she didn't realize she wasn't wearing her good wig.

Miss Kaufman was asking Mama if she knew about the child murders happening around the city. Of course Mama knew. Everybody did. All those little black boys and girls who'd left their homes to go to the corner store or the movies, only to turn up dead, dumped in ditches or in the woods. The Atlanta child murders had been in the news for more than a year. There were seventeen missing children, and fourteen had already been found dead. It was so bad that, the week before Miss Kauffman showed up at our door, the mayor had announced a curfew. Everybody under the age of fifteen had to be inside by 7 P.M.

"Yes, ma'am," Mama said, nodding her head. "Some devil out here killing our babies. It's a terrible thing."

Miss Kaufman looked down at a slim white notepad she held in her hands. "Miss Williams, it's my understanding that two of your children broke curfew last night. Is that correct?"

I could feel my eyes grow wide. It *was* true! Sweetie and Dre had come home the night before in the back of a police cruiser, with a couple of stolen bicycles sticking out of the trunk. I thought for sure they'd been arrested for stealing, but when the cop carried the bikes up to the porch, all he said to Mama was "You need to keep the children inside. It's for their own safety." Mama told the officer, "Thank you." But the minute he drove away, she beat Sweetie and Dre from the rooty to the tooty for bringing the gotdamn police to her door.

On the porch, Miss Kaufman was telling Mama that her kids were the first ones *in the whole city* to break curfew, and would Mama be willing to talk about it for the evening news. She smiled at Mama, reached over, and touched her arm. "It must be so very challenging to raise children during this difficult time," she said. "I'd like to hear *your* story."

Behind Miss Kaufman, leaning up against a white van with the *Action News* logo painted on the side, was a heavyset man with a TV camera on his shoulder. Mama's eyes shot to the cameraman, then back to Miss Kaufman.

"Just give me a minute to freshen up," Mama said, gently closing the door. "I'll be right back."

I don't think I've ever seen Mama move so fast. She ran to the kitchen and grabbed her fake teeth off the counter, then ran to the bedroom to put on her curly wig. She pulled her multicolored woolen shawl with the fringes from the back of the sofa and wrapped it around her shoulders. "How I look?" she asked, standing before me.

"Real good!" I told her.

Mama pushed open the door and stepped into the sunlight to tell her story to the world.

"To tell you the truth, Miss Kaufman," Mama said, looking directly into the camera, "I had sent my two children to the store. I told them, 'Y'all make sure you come home by curfew. There's a killer on the loose!'" She turned to Miss Kaufman. "I take good care of my babies. But you know how kids be, wanting to be out in the world like they grown. I'm just tryna do my best."

When the interview was over, Mama ran back inside and

straight to the black rotary phone hanging on the kitchen wall. She dialed Aunt Vanessa's number. "Girl," she said breathlessly, "I'm gonna be on TV!"

I used to wonder why Mama was so excited for folks to see she lost track of her own damn kids. But I came to think that it didn't matter what Miss Kaufman was asking. All that mattered was that somebody was paying attention. As long as she lived, nobody except caseworkers and police had ever asked Mama about her life. Nobody gave a damn. But Miss Kaufman leaned in close and really listened to Mama. For a moment, my mother was important.

When most folks think about the problems of growing up in the hood, they think about what it must feel like to be poor, or hungry, or to have your lights cut off. The struggle nobody talks about is what it feels like to be *invisible*, or to know in your heart that nobody cares. Mama didn't want to be famous; she wanted to be seen. All those years I thought we were so different, but when I stepped onstage and saw all those facing smiling back at me, I realized Mama and I craved the same thing.

I TOOK THE STAGE NAME "MS. PAT," which is what my kids' friends called me, and started hitting open mics around Atlanta. I didn't plan my material; I would just get up onstage and run my mouth. If something got a laugh, I'd use it again.

A local comic, Double D, saw me perform and asked me if I would drive him to his gigs out of town. In exchange, I'd get ten minutes of stage time. Double D was working the Chitlin' Circuit, which basically means everybody in the place—from the comics and the audience to the doormen—is black. Usually, these are one-night shows at a mainstream comedy club, a hotel lounge, or in a bar. Some folks call them "black night," or the "urban show." But I've heard it called other things, too.

Don't get me wrong, I love my people, but those rooms are *not* easy. You gotta come out of the gate swinging. If you don't hit the crowd with a joke in the first thirty seconds, they'll boo your ass right off stage. I once saw a doorman damn near choke out a comic for not being funny. One club I played gave the audience Nerf balls to throw at the stage if the jokes weren't good. I got hit right in the middle of my forehead.

To avoid the humiliation, I quickly figured out that if I strung together a bunch of one-liners, mostly about sex, I could get some good laughs. "I told my husband I'd suck his dick" is how I started one joke. "But I told him, 'You gotta dip that shit in some blue-cheese dressing first.'"

Michael didn't know I was making jokes about our sex life. While I was out at the clubs, he was at home watching CNN. He thought comedy was a phase I was going through. That's because I didn't tell him about my plan to make comedy my career. I reasoned if Sinbad could earn a living telling jokes, why couldn't I? Never mind that Sinbad was selling out arenas and I was telling dick jokes to a room full of people who would throw their car keys at me if I didn't make them laugh.

THEY SAY IT TAKES at least ten years to really "find your voice" as a comic. The first time I heard that expression I was confused as hell. I thought it meant my actual *voice*. "My neighbor says I sound like Moms Mabley," I told Double D. But he explained that "voice" is the thing that makes a comic different from everybody else. "You know," he said. "Your point of view."

I didn't find my voice in Atlanta, or while I was working the Chitlin' Circuit. It wasn't until 2006, when the General Motors plant in Atlanta closed. Michael's job got relocated, and we had to pick up and move to Plainfield, Indiana, a little town outside Indianapolis. That's when I finally figured it out.

I'd never been to a place as white as Plainfield. I was used to city living and black folks. Suddenly I was surrounded by people who looked like the first thing they did when they rolled out of bed in the morning was go milk a cow. I was worried about how my kids would fit in at schools where they would be the only black kids in the class. It turned out Junebug was fine. He got himself some little white friends named Ethan and Conner who didn't seem to notice the color of his skin. Nikia, a high school senior, got called "nigger" his first week of school and had to kick a white boy's ass to straighten him out. Michaela was eight years old when we moved. She was sensitive and would lay in the cut, watching and listening, cataloging a list of insults and offenses that she'd hold against those kids until the day she graduated from high school.

Out of everybody in the family, the person who had to make the biggest adjustment to white suburban life was me. I'd never told anyone, not even Michael, about the number-one lesson Mama taught me about white folks. "They better than you," she said one afternoon when we were home watching *The Price Is Right*. I was ten years old. "Remember what I'm telling you, girl. White folks is better than you. Make sure you never look them in the eye."

It wasn't the first time Mama had given me this particular piece of slavery-times advice. And it stuck with me, growing into a fear I carried all the way to Plainfield. Even though we moved into a nice house in the suburbs with ducks swimming in a little pond outside our front door, I felt uncomfortable every time I left home. Everywhere I looked was a white face smiling back at me. I was so intimidated it was hard for me to speak. A big redheaded guy once stopped me in a grocery store and mentioned that he thought our boys were on the same football team. I could barely fix my mouth to tell him yes. All I could think of was how stupid I must sound. It was all I could do to nod and look away.

That's why it took every ounce of courage I had for me to walk into Morty's Comedy Joint, on the north side of town. Morty's wasn't like the urban rooms I was used to playing where comics had to fight for attention with the TV blasting full volume behind the bar. At Morty's, the audience came ready to listen. And they were patient, even with comics who liked to take their time. Like, let's say you wanted to tell a joke about going to the drugstore. At Morty's you could walk the audience up and down the drugstore aisles, riffing about Dulcolax, stupid-ass greeting cards, and how come they don't make size "extra small" condoms, before you dropped the punch line. At an urban club, a comic would hit the stage and yell, "I told my girl, 'If you bleeding you better call 911 because I ain't buying no muthafuckin' tampons!'" and that was the whole damn joke.

At Morty's I'd hang out by the bar watching other comics' sets and wait for a chance to go up. Sometimes Avery, the manager, would be there. He looked just like one of the Three Stooges, with a round face and bald head. Pretty soon, we got to talking. I told him how I got pregnant at thirteen, got paid selling crack, and dropped five thousand dollars on a custom paint job for my Cadillac. One night Avery said, "Why are you back here telling me all this? You need to put these stories onstage."

I didn't think a bunch of middle-class white folks would relate to my life, but Avery kept insisting. One night I finally decided to give it a shot.

"Hi, y'all," I said, looking out into the audience. "I'm a mom. How many of y'all are parents? I had my kids early. Any of you guys had your kids early, too? Like fifth or sixth grade?" I told jokes about being a teenage mother and struggling to survive. I talked about Mama's baptism hustle and the way she'd fire her pistol in the house. "I wish I was lying," I said, "but this shit is true."

I couldn't believe it. Not only did I get some good laughs; the more personal I was, the more folks connected. One night after a show a blond lady carrying a Louis Vuitton purse and wearing big-ass diamond earrings came up to me and leaned in close. "I've been through what you went though," she whispered.

"Your mama made you pickpocket from drunks?" I asked.

"No. I got pregnant when I was thirteen," she said. "Those child-molesting assholes are *everywhere*."

All those times Mama told me white folks were better than

me had me thinking white people all lived the easy life. But that woman isn't the only one who's come up to me after a show to tell me about her shitty childhood or her drug-addicted parents. I was a grown woman before I found out black folks aren't the only ones who have hard times. Everybody's got a struggle. Nobody gets through this life easy.

IT TURNS OUT COMEDY AND SELLING DRUGS have a lot in common. You need to be quick, work hard, and give people what they want. But to make it in comedy, you also need a break. Back in the day, a comic could go on *The Tonight Show* and their career would take off overnight. But the industry changed. By the 2010s, it was all about podcasts. Every comic had one, and everyone wanted to be a guest. Except me. I didn't know shit about podcasting, until the summer of 2014 when comedian Eddie Ifft was looking for a guest and a friend of his, a comic from Indianapolis, recommended me. Eddie and I spent an hour talking shit about cops and drugs and black people. I guess word got out that I had some good stories to tell, because I started getting invites from all over: comics Ari Shaffir, Bert Kreischer, Joey "Coco" Diaz, Tom Segura, and Christina Pazsitzky all had me on their podcasts. Even Joe Rogan, who is famous for keeping guests on for three hours talking about outer space and hallucinogenic drugs, invited me on his show. Sometimes I felt like these white boys were using me as a ghetto tour guide so they could learn about a place they'd be too scared to visit in real life. But other times, it felt like I was opening their eyes to a reality they needed to see.

I'd been doing the podcast circuit a few months when I got a message on Twitter that changed everything. It was from Marc Maron, whose WTF podcast is one of the biggest in the game. "My fans have been asking for you," he said. Marc wanted to know if I'd come out to Highland Park, California, and do his show.

I was nervous as hell when I walked into Marc's garage, where he records his podcast. I'd heard that Louis C.K. had cried during an interview with Marc, I was worried about what he was going to ask me. Then I looked around. The place was filled with books, coffee cups, weird little knickknacks, and framed photographs and drawings covering every wall. It's hard to feel intimidated in a place that looks like the home-furnishing section at the Goodwill.

"This is nice," I said, sitting down.

Marc asked me where I grew up and the next thing I knew I was telling him about Derrick, Mama, and Granddaddy's bootleg house. I even told him about calling President Jimmy Carter the "N" word and giving him a cheeseburger for free. "I wonder if he remembers me," I said to Marc.

"I would hope so," he said. "If anything, for the free cheeseburger."

When the podcast aired that October, I noticed Marc had recorded an introduction. He said that talking to me, "learning about what it means to grow up poor and black in America," blew his mind.

I appreciate Marc so much for having me on as a guest. It helped me get a deal to write this book. But I don't know if I want to be the poster child for growing up in the hood. Not everybody has it as bad as I did. Plenty of poor black girls don't get knocked up by married-man predators, and not every kid has a mama who looks the other way. There are lots of poor folks who work hard and take care of their babies. There are teenage moms who make it out of the hood without ever selling drugs or dropping out of school. I just had the extra bad luck to be born into a family that had been beat down for so long, all that was left to our name was a bunch of hustlers and addicts. I had no one to show me the way.

I could easily have turned out different, ending up like my sister, or Butterfly, or all the other girls who I saw get lost to the streets. Instead, I feel like I was specially blessed. How else can I explain the angels who seemed to come out of nowhere when I needed them most?

Granddaddy and Curtis were the first angels who showed me love. But I was lucky, I had a whole crew. Miss Troup, my angel in badass leather boots, taught me to dream, Duck told me to act right, and Lamont opened my eyes to quality. Miss Munroe and Miss Campbell gave me good guidance and Hubert Hood watched over my kids. And through it all my children, my angel babies, made sure I never, *ever, ever* gave up.

Of all the angels I had looking out for me, Michael is the boss. He met me when I was struggling and scheming. But he saw the good in me and believed I could do better. He wrapped his arms around me and didn't let me go.

People ask me all the time how I turned my life around. I used to think it was too complicated to answer without telling my whole life story. But now that I've laid it all out in black and white, I realize the answer is really pretty simple: I wanted to turn my life around, and what got me there was love.

EPILOGUE

December 26, 2013

The place I was looking for was in a part of Decatur, Georgia, I'd never been to before, in the kind of neighborhood I hadn't driven through in more than a decade. There were houses with busted-out windows and run-down buildings with beat-up sofas and empty soda cans littering their weed-filled yards.

"We in the hood now," I said to Nikia, who was riding with me, as we scanned the street.

We were looking for an address I'd put in my car's GPS, searching for Sweetie's second daughter, Diamond, who I hadn't seen in years. Her older sister, LaDontay, had contacted me saying Diamond needed help, and asking could I go check on her. LaDontay didn't call their mama; she called me.

"I don't know about this," I said to Nikia.

"Everybody deserves a chance," he said.

WE FOUND DIAMOND IN AN upstairs apartment that smelled like dog piss, dirty feet, and weed. Dressed in faded pink sweatpants and with nappy hair, she was standing in the middle of the living room looking dazed. Peeking out from behind her legs was her six-year-old son, Jamal, dressed in a thin windbreaker and win-ter boots several sizes too large. His younger sisters, aged two and four, clung to Jamal's bony frame. Cradled in Diamond's arms, wrapped in a hospital blanket, was her two-week-old baby girl.

Diamond told me she had been waiting on her child's father to come get her. "But he got locked up," she said. Now the people she was staying with—her boyfriend's cousin's family—wanted her out.

A woman shuffled into the living room and pulled me aside: "Real talk, that nigga she waiting on ain't *never* gettin' out. She needs to get up outta here and go."

I hadn't seen Diamond since before Michael and I moved to Indiana, before her son was born. I didn't know what hardships she'd faced or what choices she'd made or which road she'd trav-eled down that led her to this particular corner of hell. All I knew is she'd hit rock bottom, and now the bottom was falling out.

I reached out my hand for her little boy and told Diamond to gather up her things. As we carried her one suitcase and a trash bag of possessions to the car, I started rehearsing in my head how I was gonna tell Michael that I was bringing home five more mouths to feed.

DIAMOND CAME TO LIVE WITH US IN INDIANA. I took her to a dentist, who pulled out sixteen of her teeth because they were rot-ting in her head. And I arranged for her to get some glasses be-cause she could barely see. I helped her get day care for her girls and enroll Jamal in school. One night before bed, I heard him ex-plain to his little sisters in a serious tone, "At Auntie Pat's we get to eat *every single day.*"

 Diamond got herself a job working the night shift at a bakery, cleaning the giant machines where they make the bread. Then, two years after I'd picked her up in that shit hole of an apartment out in Decatur, she'd saved up enough money to move herself and her kids into their very own place. I was so proud. I told her every day, "Girl, you work hard you can turn your life around!" Michael and I even cosigned so she could get herself a car.

A few months later, she crashed it.

Then she fell behind in her day-care bill and stopped paying her rent. She didn't tell me any of this. Instead, she just disappeared.

I don't know exactly where she is now. I saw a picture of her on Facebook a few weeks ago. She was in the backseat of a car, her head rolled to the side, her eyes barely open, wearing a bright blond wig and the glasses I'd gotten her. I guess whatever demon she's fighting won this battle. But I'm not giving up on Diamond. I raised that girl for ten years. I know she can find her way back.

In the meantime, her kids are living with me and Michael in the suburbs. Michaela and Junebug watch over them after school and Michael reads to them every night.

"I love you, babies," I said this morning while getting them dressed for school. "And I want you to know something important. No matter what kind of hard times you face, remember you can do anything and be anything you want in life."

I pulled them to me and held them close, my mind thinking back to all the angels who'd kept me safe, given me hope, and helped me find my way. "Remember," I whispered, "All you have to do is dream."

HOW THIS BOOK CAME TO BE

I first heard Ms. Pat when she was a guest on Ari Shaffir's *Skeptic Tank* podcast in June 2014. I remember walking my dog and listening to Pat's nonstop laughter as she described becoming a mother, selling crack, and getting shot in the head, all before her seventeenth birthday.

It wasn't the first time I'd heard details like this. For twenty years I've worked as a journalist. I've interviewed drug dealers, hustlers, and young girls who've been beaten, shot at, and impregnated by men almost twice their age. But never have I talked to someone who embodied all these issues at once.

A few nights later I went to see Pat perform at Union Hall, a little club near where I live in Brooklyn. I brought my friend Sharon along. After the show Sharon ran up to Pat and exclaimed, "Your story's crazy. You should write a book!"

Pat said, "I've always wanted to, but I'm not a writer."

Sharon pulled me over: "She can help you write your book."

A few weeks later I called Pat. We ended up talking for hours. Pat's journey from illegal liquor house to the suburbs of Indianapolis isn't just incredible. It's also the type of story too few people get to hear. Popular culture has given us plenty of depictions of boys in the hood. But what about the girls? What do most people know about the challenges of being poor, black, and female? Not much. Instead, young black mothers living in poverty are often described as "irresponsible" and "lazy." I've even heard them called "animals." It pains me to hear young women so callously dismissed by people who don't know their lives. I could see in Pat a unique opportunity to help bring one of these stories to light. By the end of our conversation, I'd committed to helping her write this book.

I assumed we'd collaborate the way most subjects and writers do: I'd interview Pat, get all the details, and put the story together in a way that makes sense on the page. But Pat is one of those people blessed with an extraordinary memory. When I asked her to describe her grandfather's liquor house, she could recall every detail right down to the pattern on the faded bedsheets hanging in the windows. This would have been a great advantage had Pat's life been less eventful. Instead, often I'd pose a simple question like, "What color did you paint your Cadillac?" Forty-five minutes later I'd find myself stunned by some new revelation Pat had casually tossed into the conversation.

Every moment of Pat's life is a movie unto itself, and she remembers it all in vivid detail. Deciding which anecdotes *not* to include became one of the greatest challenges of this book. In the end, I focused on telling the story Pat most wanted to share, about how she made it out of the hood and turned her life around when all the chips were stacked against her. I hope I did her journey justice.

During the more than two years we spent working on this book, I spoke to almost a dozen people from Pat's life, including "Duck," "Stephanie," and her siblings "Dre" and "Sweetie." I interviewed Pat's caseworker "Miss Campbell," and managed to locate the criminal lawyer who defended her more than two decades ago. I tracked down "Hubert" Hood, whom Pat hadn't spoken to in years, and talked to Pat's cousin "Tata," her children, and her husband, Michael.

Sometimes the interviews were scheduled in advance; other times the conversations would happen spontaneously, when I least expected it. Pat would be in the middle of a story, and I'd ask a question about the particulars of selling crack. The next thing I knew she'd tell me to hang on and suddenly some guy she used to buy crack from would be on the line.

"This the lady writing my book," she'd say by way of introduction. And then to me, "Go ahead, ask your question."

The conversations didn't always go smoothly. Occasionally I'd ask for clarification only to be met with an uncomfortable silence. In those instances, Pat would intervene with her magic words, "It's okay, you can talk to her. She's black. She just *sounds* white."

I studied Pat's criminal history record, called courthouses and county jails. I researched the progress of crack through inner cities in the late eighties and nineties, and the war on drugs waged by the nation in general and the Atlanta Police Department in particular. I studied the history of the city's housing projects and the horrific Atlanta Child Murders. I discovered that one of the victims, Curtis Walker, was Pat's second cousin.

Of course there were stories Pat told me for which I had to rely solely on her telling, but for everything I checked and cross-checked, not once did I find an inconsistency. In fact, when Pat shared her experiences of something that had been reported in the news—for instance, the Miami Boys' infiltration of Atlanta housing projects in the late 1980s—her recollections aligned, right down to the little details, with decades-old newspaper accounts I was able to dig up.

EVERYBODY LOVES PAT'S STORY. I've heard her on countless podcasts with hosts who marvel both at the hellish world in which she was raised and her determination to get out. But here's the thing, as shocking as the details of Pat's life are, the circumstances that led her to sell crack to survive are not all that unique. A few months into writing this book I took a break to work on another assignment, which involved my interviewing economically disadvantaged teen mothers around the country. I spoke with scores of girls who, neglected or abandoned by their own parents, were impregnated by much older men. I learned of young mothers so poor they were living with their infant children in homes with no heat or running water in the middle of winter. I interviewed girls whose greatest challenge was finding diaper money even as they dreamed of a better life for their kids. If you think Pat's story couldn't happen today, you're wrong. It's happening still.

There is a myth in this country that the way out of poverty is to "pull yourself up by your bootstraps," that by sheer force of will one can change the course of one's life, no matter how great the obstacles. But in all my years reporting, I've never once spoken to someone who came from abject poverty and transcended that path without help. Pat talks about how badly she wanted to change her life, but she couldn't have made that dream a reality without the compassion and care she received from Miss Troup, "Hubert" Hood, "Lamont," and, of course, "Michael." She couldn't have done it without public assistance and the support provided by her caseworkers "Miss Campbell" and "Miss Munroe." Pat wanted to share her journey to inspire people to get over their own hardships. But the more we talked, the more clear it became: Pat's story is greater than a tale of one woman rising above her own harrowing circumstances. It's a testament to the transformative power of love.

—*Jeannine*

ACKNOWLEDGMENTS

From Ms. Pat

To my fans, thank you for all your support. It means everything to me. For years I wanted to tell my story but I didn't think anyone would ever be able to relate to the things I've gone through in my life. But the more I talked, the more you guys supported me. I thank you from the bottom of my heart and encourage you to tell your stories and laugh when you feel like crying.

Thank you to my husband, Garrett, for being by my side when everyone else left me. I thank God every day for answering my prayers and sending you my way. God blessed me with everything I asked for in a man: you're good looking, smart, a good provider, and you have all your back teeth.

Thank you to my wonderful kids who gave me the will to survive: Ashley, thank you for never judging me and understanding that we were growing together. Thanks to my second-born child, Nikia: you have one of the biggest hearts of anyone I've ever met. Also thank you for blessing me with three beautiful grandchildren. Thank you, Garrianna, for always making me laugh. And to my last-born, Garrett Jr., thank you for going undetected for six months so my only choice was to have you. You were ten pounds of joy! Kids are supposed to make parents proud. Thank you all for always telling me how proud you guys are of *me*.

Thank you to my cousin Boo. I'm sorry for the things I exposed you to at such a young age. I wanted to provide a better life for you and ended up keeping you from living a normal childhood. If I could give you back the childhood you deserved, I would. I hope one day you find the strength to forgive me and know I love you.

Thank you Ant, Tony, and Bo for being the best big brothers I could ever ask for (now that we're grown!). And to my sister, May-pop, I know you won't be able to read this because it's in proper English. I'm just kidding! I'm sorry for what we had to go through as kids. You know I love you.

Thank you to Jasper Hood for keeping my kids safe and keeping your business open a little bit longer so they could stay warm. You were like a father figure to me when I needed one the most. Thank you Miss Chambers for laughing so damn hard when you were my welfare caseworker and for being the mother figure I never had. And thank you, Miss Troup: you showed me kindness at a time in my life when no one else cared. Rest in peace.

To my girlfriends Melodie and Lisa: for thirty years you've both been there when I needed you, giving me good advice even when I was too stubborn to take it and making me laugh when I thought all I could do was cry. Thank you, Tracy, for accepting me for who I am, even though I'm a girl from the hood and you have five degrees. And Ms. Jeanne, thank you for opening up your life and letting my crazy ass into it. Friendship is the most pure and honest form of love. Thank you all.

To my loving manager, John, who came into my life like a step-daddy and turned my career around. Thank you for being so damn honest with me, and not leaving me when I didn't pay you. Thank you for always respecting me, never judging me, keeping me calm, and teaching me to be patient.

Thank you to my podcast family for helping me get my stories out there: Eddie Ifft, Joey Diaz, Bert Kreischer, Ryan Sickler, Jay Larson, Joe Rogan, Tom Segura, Christina Pazsitzky, Moshe Kasher, and Neal Brennan. To Marc Maron, thank you for inviting me onto your podcast and allowing an E-list celebrity to tell her story. And to Ari Shaffir, you seemed like an asshole when I first met you, but as I got to know you I realized you were just a dick, and I like dick.

To all the comics who were there for me when I was first starting out: Katt Williams, Double D, Arnez J., and D Ray, you all gave me opportunities, taught me the business, and showed me respect. Thank to you to Kourtlyn Wiggins, for fifteen years of always keeping it real with me and spending all that time on the phone discussing jokes and life. And thank you to your wife for putting up with the late-night calls. Thank you to Avery from Morty's Comedy Joint for encouraging me to put my stories onstage, and to the Bob and Tom radio show for all your support.

Thank you, Brandi Bowles from Foundry Literary + Media, for keeping the wheels of this double-decker bus rolling and always steering us in the right direction. And to Julia Cheiffetz at HarperCollins, thank you for believing in my story and helping to make this the best book it could be. And thank you for the book money, it really helped me pay off my layaways.

Last but not least, thank you, Jeannine, for writing this book. I told you things I've never told anybody before. It wasn't easy, but talking to you helped me heal. So thank you for crying with me and laughing with me and never judging me. Thank you for listening to me even when I was mean. I know I drove you crazy, but I appreciate all the super intelligent arguments we had (that I always let you win). Thank you for being my friend.

From Jeannine

First and foremost, thank you, Pat, for trusting me with your story and sharing not only the details in these pages, but also the funny tangents and painful moments you shared only with me. I'd also like to say a special thanks for that time you yelled at me when I asked you describe in more detail exactly how to break down a quarter ounce of crack. "Jeannine!" you shouted. "Why do you keep asking me this shit? All you do is chop it up!" Your exasperation at my failing to possess this highly specialized information, which apparently you think everyone just knows, is my personal comedy highlight of this entire project. Pat, you and I spent hours talking, laughing, crying, and occasionally driving each other crazy. I've gained a lifelong friend.

To my agent Brandi Bowles at Foundry Literary + Media, thank you a million times over for your guidance, encouragement, and sound advice. This book couldn't have happened without you. To my editor Julia Cheiffetz, at HarperCollins, thank you for believing in this project and pushing me to find moments of levity to balance out the heartbreak. This was a long journey, and Pat and I appreciate your being with us every step of the way.

To Marc Maron, your amazing interview with Pat (*WTF*, episode 540) caught Julia's attention just as the book proposal hit her desk. What a gift! Thank you, Marc, for changing lives one podcast at a time. Thank you also to Ari Shaffir for introducing me to Pat with your hilarious interview (*Skeptic Tank,* episode 169).

To my brilliant friends Kisha, Rosie, and Randi, thank you for all your support, feedback, and insight. You didn't just help me get through the arduous process of writing a book; you also helped me get through (the more arduous and less lucrative) process of life. To Eva (and your mom), thank you for the southern-isms. How else would I know about Cooter Brown? Thanks to Sharon, who insisted Pat and I made a good match. And to Stan, for your kind generosity.

Most of all, thank you to my family. Sometimes I can't tell if a chapter is working unless I read it out loud. Mom and Dad, thank you for never saying no when I asked you to listen. You are unwavering in your support, limitless with your love, and the best parents a writer could have. To my sibs, Gillian and David, thanks for everything. And to my young niece and nephews, Jacqueline, Harrison, and Ethan, please don't tell your parents that when I said we were all "going to play Monopoly" that I was really reading you kids a chapter detailing the complexities of crack distribution in the hood. Your enthusiastic response was a little unsettling, but thanks for listening. Last, to my daughter Niko, thank you for understanding. I love you most of all.

ABOUT THE AUTHORS

PATRICIA "MS. PAT" WILLIAMS is a stand-up comedian, actress, and writer. She broke into the comedy industry in Atlanta at the notorious Uptown Comedy Club, and now headlines at clubs throughout the country. Ms. Pat has appeared on NBC's *Last Comic Standing,* Comedy Central's *This Is Not Happening* Nickelodeon's *NickMom Night Out,* Katt Williams's DVD release *Kattpacalypse,* Harry Connick Junior's syndicated talk show, *Harry,* and numerous podcasts, including Marc Maron's WTF, The Joe Rogan Experience, and Ari Shaffir's Skeptic Tank. She is a regular guest on the syndicated radio program *The Bob & Tom Show.* Ms. Pat currently resides in Indianapolis with her family.

JEANNINE AMBER is an award-winning journalist and former senior writer for *Essence* magazine. For two decades she's written about crime, relationships, and parenting. And also . . . Beyoncé! Jeannine loves a good story. Check out her work at jeannineamber .com or follow her on Twitter @jamberstar.